CAMBRIDGE MONOGRAPHS ON MATHEMATICAL PHYSICS

KINETIC THEORY IN THE EXPANDING UNIVERSE

KINETIC THEORY IN THE EXPANDING UNIVERSE

JEREMY BERNSTEIN
Stevens Institute of Technology and Rockefeller University

CAMBRIDGE UNIVERSITY PRESS
Cambridge
New York New Rochelle Melbourne Sydney

PUBLISHED BY THE PRESS SYNDICATE OF THE UNIVERSITY OF CAMBRIDGE
The Pitt Building, Trumpington Street, Cambridge, United Kingdom

CAMBRIDGE UNIVERSITY PRESS
The Edinburgh Building, Cambridge CB2 2RU, UK
40 West 20th Street, New York NY 10011–4211, USA
477 Williamstown Road, Port Melbourne, VIC 3207, Australia
Ruiz de Alarcón 13, 28014 Madrid, Spain
Dock House, The Waterfront, Cape Town 8001, South Africa

http://www.cambridge.org

First published 1988
First paperback edition 2004

A catalogue record for this book is available from the British Library

Library of Congress Cataloguing-in-Publication Data
Bernstein, Jeremy, 1929–
Kinetic theory in the expanding universe/Jeremy Bernstein.
p. cm. – (Cambridge monographs on mathematical physics)
Bibliography: p.
Includes index.
ISBN 0 521 36050 1 hardback
1. Expanding universe. 2. Matter, Kinetic theory of.
3. Transport theory. 4. General relativity (Physics)
5. Astrophysics. I. Title. II. Series.
QB991.E94B47 1988
523.1′8 – dc 19 88–4250

ISBN 0 521 36050 1 hardback
ISBN 0 521 60746 9 paperback

Transferred to digital printing 2003

Contents

Preface

The purpose of this monograph is to provide a bridge between "two cultures." On one side are mathematically inclined relativists such as J. M. Stewart (Stewart, 1971) and J. Ehlers (Ehlers, 1971), who have constructed the formal mathematics for kinetic theory in general relativity and on the other, the practical cosmologists who are concerned with physically interesting results obtained with a minimum of formalism. A partial bridge between these communities was erected by Weinberg (Weinberg, 1971, 1972). However, Weinberg's discussion focuses on macrophysics, ignoring for the most part the underlying microphysics. Our discussion will emphasize the microphysics. We will therefore be concerned with the solutions of the Boltzmann equations in the expanding universe for various cosmological processes. We will derive these equations from first principles and solve them as explicitly as possible. Many of the results will be familiar to practicing cosmologists but are presented here with more pedagogical rigor than is generally found in the literature. The treatment is meant to be essentially self-contained, but someone not familiar with the cosmological implications of general relativity would be advised to consult a book like that of Weinberg (Weinberg, 1972).

Now, by way of acknowledgment: My interest in this subject was first aroused by a series of informal lectures given by my Stevens Institute colleague J. L. Anderson (Anderson, 1970). I recall how surprised I was to learn that a Robertson–Walker universe did not, in general, admit equilibrium solutions to the Boltzmann equations (details are given later in the book). I, in turn, managed to persuade my friends and colleagues Lowell Brown and Gary Feinberg of the interest of these matters and, for the past several years, we have been collaborating on them. Much of this work is being published for the first time here. The reader will often find a footnote in this book that reads, "This work was done in collaboration with L. Brown and G. Feinberg," and so it was. The book was written over several summers at the Aspen Center for Physics. My gratitude to the Physics Center is very great indeed. I am also very grateful to Erick Weinberg,

Scott Dodelson, Subir Sarkar and Larry Widrow for their critical reading of the manuscript and many suggestions, and to Rufus Neal of the Cambridge University Press for both his encouragement and careful editing of the manuscript. Finally, I would like to thank Joan Leatherbury for typing a very complicated manuscript with cheerful goodwill.

New York, 1988 Jeremy Bernstein

1
Tools: relativity

In this section we introduce some results from the general theory of relativity. More details can be found, for example, in Weinberg (1972).

The fundamental quantity in the theory is the metric tensor $g_{\mu\nu}$. In fixing it we adopt the usual point of view in cosmology that local metric irregularities due to stars and galaxies, and the like, are ignorable in the large. This is canonized into what is known as the "cosmological principle"; i.e., that the universe, taken on average, is and always has been homogeneous and isotropic. That is to say, at any epoch the universe appears the same in all spatial directions when observed from any spatial point. The first person to use this principle to derive, in a mathematically satisfactory way, the form of the metric tensor was H. P. Robertson (Robertson, 1929). He showed, in units in which $\hbar = c = 1$, that

$$ds^2 \equiv -g_{\mu\nu}\,dx^\mu\,dx^\nu = dt^2 - e^{-f}h_{ij}\,dx^i\,dx^j. \tag{1.1}$$

Here the h_{ij} are functions of the spatial variables x^1, x^2, x^3 and f is an arbitrary, real, function of the time t. Robertson gives as an example the case in which

$$f = t \times \text{constant},$$

which is what we now call the de Sitter inflationary universe (de Sitter, 1917). This, as we shall shortly show, is one of the two cases in which the curvature of the universe remains constant in time. The other case is the so-called Einstein cosmology (Einstein, 1917) in which

$$f = 0.$$

Introducing polar coordinates, as Robertson did, we can rewrite (1.1) in the now familiar form

$$ds^2 = dt^2 - R^2(t)\left(\frac{dr^2}{1 - kr^2} + r^2\,d\theta^2 + r^2\sin^2\theta\,d\phi^2\right). \tag{1.2}$$

The scales of R and r can be chosen so that $k = 0, \pm 1$. Some general relativity formulae will be useful. In terms of the metric tensor, $g^{\mu\nu}$, the covariant derivatives of an arbitrary vector field A_ν or $A^\mu = g^{\mu\nu}A_\nu$, where $g^{\mu\nu}$ is given by the relation

$$g^{\mu\nu}g_{\lambda\nu} = \delta^\mu_\lambda, \tag{1.3}$$

are defined by the equations

$$A^\nu{}_{;\mu} = A^\nu{}_{,\mu} + \Gamma^\nu_{\mu\lambda}A^\lambda, \tag{1.4}$$

and

$$A_{\nu;\mu} = A_{\nu,\mu} - \Gamma^\lambda_{\mu\nu}A_\lambda, \tag{1.5}$$

where, as is customary, the comma denotes the ordinary derivative and $\Gamma^\mu_{\nu\sigma}$ is the Christoffel symbol defined in terms of the metric by

$$\Gamma^\mu_{\nu\sigma} = \tfrac{1}{2}(g_{\lambda\sigma,\nu} + g_{\nu\lambda,\sigma} - g_{\nu\sigma,\lambda})g^{\lambda\mu}. \tag{1.6}$$

For the metric of (1.2) the $g_{\mu\nu}$ are given by

$$g_{00} = -1, \tag{1.7}$$

$$g_{i0} = 0,$$

and

$$g_{ij} = R^2(t)\hat{g}_{ij}, \tag{1.8}$$

where, in polar coordinates,

$$\hat{g}_{rr} = \frac{1}{1 - kr^2}, \tag{1.9}$$

$$\hat{g}_{\theta\theta} = r^2,$$

$$\hat{g}_{\phi\phi} = r^2 \sin^2\theta,$$

$$\hat{g}_{ij} = 0, \quad i \neq j.$$

We are implicitly using here what are known as "comoving" coordinates. This means that a galaxy, say, is assigned values of r, θ, and ϕ which remain the same as the universe expands. The points on the mesh which define the coordinate grid, expand with the grid. If we call

$$g = -\det g_{\mu\nu} \tag{1.10}$$

then, for our metric

$$g = \frac{R^6 r^4 \sin^2\theta}{1 - kr^2}, \tag{1.11}$$

and the invariant volume element is given by

$$dV = g^{\frac{1}{2}} dx_1 \, dx_2 \, dx_3 \, dx_0. \tag{1.12}$$

Because of the simplicity of the metric the only nonvanishing Christoffel components are

$$\begin{aligned}
\Gamma^0_{i\,j} &= R\dot{R}\hat{g}_{ij} \\
\Gamma^i_{0j} &= \frac{\dot{R}}{R}\,\delta^i_j = \Gamma^i_{j0} \\
\Gamma^1_{11} &= \frac{kr}{1-kr^2} \\
\Gamma^1_{22} &= -(1-kr^2)r \\
\Gamma^1_{33} &= -(1-kr^2)r\sin^2\theta \\
\Gamma^2_{12} &= \Gamma^2_{21} = \Gamma^3_{13} = \Gamma^3_{31} = 1/r \\
\Gamma^2_{33} &= -\sin\theta\cos\theta \\
\Gamma^3_{23} &= \Gamma^3_{32} = \cot\theta.
\end{aligned} \tag{1.13}$$

In most of our later work we shall restrict ourselves to the $k = 0$ (spatially flat) case. If we write the spatially flat metric as

$$ds^2 = dt^2 - R^2(t)(dx_1^2 + dx_2^2 + dx_3^2), \tag{1.14}$$

we have, with this Cartesian choice,

$$\begin{aligned}
g_{00} &= -1 \\
g_{i0} &= 0 \\
g_{ij} &= R^2(t)\delta_{ij},
\end{aligned} \tag{1.15}$$

and

$$\begin{aligned}
\Gamma^0_{ij} &= R\dot{R}\delta_{ij} \\
\Gamma^i_{0j} &= \dot{R}\delta^i_j/R \\
\Gamma^i_{jk} &= 0.
\end{aligned} \tag{1.16}$$

Because of the rotational symmetry of the sections of space orthogonal to a given time direction the spatial components of the Riemann–Christoffel tensor $R_{\alpha\beta\gamma\delta}$ take this form in the $k \neq 0$ case:

$$R_{ijkl} = \Lambda(g_{jk}g_{il} - g_{jl}g_{ik}). \tag{1.17}$$

This satisfies the general symmetry conditions

$$R_{\alpha\beta\nu\delta} = -R_{\alpha\beta\delta\nu} = -R_{\beta\alpha\nu\delta} = R_{\nu\delta\beta\alpha} = R_{\beta\alpha\delta\nu}. \tag{1.18}$$

In (1.17) Λ is a time-dependent spatial scalar. To evaluate it we use the general definition of $R_{\alpha\beta\nu\delta}$

$$R_{\alpha\beta\nu\delta} = \frac{1}{2}\left(\frac{\partial^2 g_{\alpha\beta}}{\partial x^\nu \partial x^\delta} - \frac{\partial^2 g_{\beta\nu}}{\partial x^\alpha \partial x^\delta} - \frac{\partial^2 g_{\alpha\delta}}{\partial x^\beta \partial x^\nu} + \frac{\partial^2 g_{\beta\delta}}{\partial x^\alpha \partial x^\nu}\right) + g_{\eta\sigma}(\Gamma^\eta_{\nu\alpha}\Gamma^\sigma_{\beta\delta} - \Gamma^\eta_{\delta\alpha}\Gamma^\sigma_{\beta\nu}). \tag{1.19}$$

Because of the structure of (1.17) any spatial component will do. Thus, for example,

$$R_{1221} = \frac{r^2 R^2}{1 - kr^2}(k + \dot{R}^2), \tag{1.20}$$

from which it follows that

$$\Lambda = \frac{k + \dot{R}^2}{R^2}. \tag{1.21}$$

The Ricci tensor is defined to be

$$\begin{aligned} R_{\mu\kappa} &= g^{\lambda\nu} R_{\lambda\mu\nu\kappa} = R_{\kappa\mu} \\ &= g^{ij} R_{i\mu j k} + g^{00} R_{0\mu 0\kappa}. \end{aligned} \tag{1.22}$$

We begin by evaluating its spatial components. Thus

$$\begin{aligned} R_{lm} &= g^{ij} R_{iljm} + g^{00} R_{0l0m} \\ &= g^{ij}\Lambda\{g_{lj}g_{im} - g_{lm}g_{ij}\} - R_{0l0m} = -2\Lambda g_{lm} - R_{0l0m} \end{aligned} \tag{1.23}$$

As

$$R_{0l0m} = \ddot{R}R\hat{g}_{lm}, \tag{1.24}$$

$$R_{lm} = -\hat{g}_{lm}[2k + 2\dot{R}^2 + \ddot{R}R].$$

On the other hand

$$\begin{aligned} R_{0\alpha} &= g^{\lambda\nu} R_{\lambda 0 \nu\alpha} \\ &= g^{ij} R_{i0j\alpha} + g^{00} R_{000\alpha}, \end{aligned} \tag{1.25}$$

from which it follows that

$$R_{00} = g^{ij} R_{i0j0} = 3\ddot{R}/R, \tag{1.26}$$

while

$$R_{0i} = 0. \tag{1.27}$$

The invariant curvature ${}^{(4)}R$ is defined to be

$$
\begin{aligned}
{}^{(4)}R &= g^{\lambda\nu} g^{\mu\kappa} R_{\lambda\mu\nu\kappa} \\
&= g^{\mu\kappa} R_{\mu\kappa} = g^{00} R_{00} + g^{ij} R_{ij} \\
&= -6\left(\frac{k}{R^2} + \frac{\dot{R}^2}{R^2} + \frac{\ddot{R}}{R}\right).
\end{aligned} \tag{1.28}
$$

The statement that this curvature remains constant in time leads, with $k = 0$, to the equation

$$\frac{\ddot{R}\dot{R}}{R} - 2\frac{\dot{R}^3}{R^2} + \dddot{R} = 0. \tag{1.29}$$

This equation, as advertised, has only the solutions

$$R = R_0 e^{at} \tag{1.30}$$

and

$$R = R_0, \tag{1.31}$$

which is to say only the Einstein and de Sitter spaces evolve with constant curvature.

The covariant divergence of a contravariant tensor is defined to be

$$T^{\mu\nu}{}_{;\mu} = \frac{1}{g^{\frac{1}{2}}} \frac{\partial}{\partial x^\mu}(g^{\frac{1}{2}} T^{\mu\nu}) + \Gamma^{\nu}_{\mu\lambda} T^{\mu\lambda}. \tag{1.32}$$

We may verify, using this definition, that

$$(R^{\mu\nu} - \tfrac{1}{2} g^{\mu\nu(4)}R)_{;\mu} = 0. \tag{1.33}$$

We have for a covariant tensor

$$T_{\mu\nu;\lambda} = \frac{\partial T_{\mu\nu}}{\partial x^\lambda} - \Gamma^{\rho}_{\mu\lambda} T_{\nu\rho} - \Gamma^{\rho}_{\nu\lambda} T_{\mu\rho} \tag{1.34}$$

or

$$T_{\mu\nu;\mu} = \frac{\partial T_{\mu\nu}}{\partial x^\mu} - \Gamma^{\rho}_{\mu\mu} T_{\nu\rho} - \Gamma^{\rho}_{\nu\mu} T_{\mu\rho}. \tag{1.35}$$

A Killing vector is a four-vector that satisfies the condition

$$A_{\mu;\nu} + A_{\nu;\mu} = 0. \tag{1.36}$$

For any component ν – no sum,

$$A_{\nu;\nu} = 0. \tag{1.37}$$

We can also write the Killing condition as

$$A_{\mu;\nu} + A_{\nu;\mu} = A_{\mu,\nu} + A_{\nu,\mu} - 2\Gamma^{\alpha}_{\mu\nu}A_{\alpha} \qquad (1.38)$$
$$= g_{\mu\rho}A^{\rho}_{,\nu} + g_{\nu\rho}A^{\rho}_{,\mu} + g_{\mu\nu,\rho}A^{\rho} = 0.$$

As we shall see, an important remark for the kinetic theory in a Robertson-Walker expanding universe is, that for nonconstant $R(t)$, there are no nonvanishing *timelike* spatially independent Killing vectors.† To this end, suppose that A^{ρ} were such a vector. We could then find a frame of reference in which the only nonvanishing component is A^0. In this frame (1.38) takes the form

$$g_{\mu 0}A^{0}_{,\nu} + g_{\nu 0}A^{0}_{,\mu} + g_{\mu\nu,0}A^{0} = 0. \qquad (1.39)$$

If we take $\mu = \nu = 1$ then

$$g_{11,0}A^{0} = \frac{1}{1 - kr^{2}}\frac{\mathrm{d}}{\mathrm{d}t}R^{2}A^{0} = 0. \qquad (1.40)$$

from which the conclusion follows.

We next turn to cosmodynamics.

† The more general statement is that for nonstationary Robertson–Walker matrixes there is no spacelike Killing vector. I am grateful to E. Weinberg for a proof of this theorem. The weaker version given in the text is sufficient for our purposes.

2
Tools: cosmodynamics

In the Robertson–Walker cosmology there is a single undetermined quantity, namely $R(t)$ – the cosmic scale factor. This is determined by the Einstein equation

$$R^{\mu\nu} - \tfrac{1}{2}g^{\mu\nu\,(4)}R = F^{\mu\nu}, \tag{2.1}$$

where $F^{\mu\nu}$ is a tensor that satisfies the divergence condition

$$F^{\mu\nu}{}_{;\mu} = 0, \tag{2.2}$$

and the symmetry condition

$$F^{\mu\nu} = F^{\nu\mu}. \tag{2.3}$$

Equation (2.1) can be rewritten by taking the trace of both sides with respect to $g_{\mu\nu}$. Thus

$$^{(4)}R = -g_{\lambda\mu}g_{\lambda\nu}F^{\mu\nu} \tag{2.4}$$

or

$$R_{\mu\nu} = F_{\mu\nu} - \tfrac{1}{2}g_{\mu\nu}g_{\lambda\alpha}g_{\lambda\beta}F^{\alpha\beta} \tag{2.5}$$

Note that the no-source term, $F^{\mu\nu} = 0$ case, yields two possible solutions – see (1.26) and (1.28); i.e.,†

(*a*) $k = 0$; $R = 0$
(*b*) $R^2 = -k$.

The latter is consistent only if $k < 0$, which means that $R \sim t$, which represents an empty universe that expands indefinitely. For a nonvanishing $F^{\mu\nu}$ we will use (2.1). This gives for the time component

$$R_{00} - \tfrac{1}{2}g_{00}{}^{(4)}R = -3\left[\frac{k}{R^2} + \left(\frac{\dot{R}}{R}\right)^2\right] = F_{00}, \tag{2.6}$$

† In suitable coordinates these can be considered part of a single solution.

which we write in the form

$$\frac{\dot{R}^2}{R^2} = -\tfrac{1}{3}F_{00} - \frac{k}{R^2}. \tag{2.7}$$

On the other hand, for the spatial components, we have

$$g_{ij}\left[\frac{k}{R^2} + \frac{\dot{R}^2}{R^2} + 2\frac{\ddot{R}}{R}\right] = F_{ij}. \tag{2.8}$$

The divergence condition (2.2) translates, before simplification, into

$$F^{\mu\nu}{}_{;\mu} = \frac{(1-kr^2)^{\frac{1}{2}}}{R^3r^2\sin\theta}\frac{\partial}{\partial x^\mu}\left[\frac{R^3r^2\sin\theta\, F^{\mu\nu}}{(1-kr^2)^{\frac{1}{2}}}\right] + \Gamma^{\nu}_{\mu\lambda}F^{\mu\lambda} = 0. \tag{2.9}$$

To make contact with the conventional cosmodynamical equations we write the components of $F_{\mu\nu}$ in the following way;

$$F_{00} = -\Lambda - 8\pi G_N\rho \tag{2.10}$$

and

$$F_{ij} = g_{ij}(\Lambda - 8\pi G_N P). \tag{2.11}$$

In these equations ρ is the positive definite energy density, Λ is the cosmological constant, and P is the pressure which, in principle, may have either sign. To justify the factor $8\pi G_N$, where G_N is Newton's gravitational constant, let us re-derive (2.7) for $\Lambda = 0$ in a manner which seems approximate but can actually be made rigorous relativistically. (Callan, Dicke, and Peebles, 1965). To this end, consider a mass-point m located on the surface of a sphere of radius $R(t)$. We assume the mass in the sphere is distributed uniformly with a constant density ρ. Because of the symmetry, the mass of the sphere acts on m as if it were concentrated at the center. This total mass is given by

$$M = \tfrac{4}{3}\pi R^3\rho. \tag{2.12}$$

Thus Newton's law reads

$$m\ddot{R} = -G_N\frac{mM}{R^2}, \tag{2.13}$$

or

$$\dot{R}\ddot{R} = -G_N M\frac{\dot{R}}{R^2}. \tag{2.14}$$

Since M is constant in time

$$\frac{\mathrm{d}}{\mathrm{d}t}(\tfrac{1}{2}\dot{R}^2) = -G_N M\frac{\dot{R}}{R^2} = \frac{\mathrm{d}}{\mathrm{d}t}\left[\frac{G_N M}{R}\right]. \tag{2.15}$$

Thus

$$\tfrac{1}{2}\dot{R}^2 - \frac{G_N M}{R} = E, \tag{2.16}$$

where E is a constant. Using (2.12) we find

$$\frac{\dot{R}^2}{R^2} - \tfrac{8}{3}\pi G_N\rho = \frac{2E}{R^2}. \tag{2.17}$$

We may now compare this with the expression derived from general relativity; i.e., setting $\Lambda = 0$ we have

$$\frac{\dot{R}^2}{R^2} = \tfrac{8}{3}\pi G_N\rho - \frac{k}{R^2}. \tag{2.18}$$

It is clear that with the identification

$$2E = -k$$

the two expressions are identical, justifying the choice of constants in (2.10). With the sign conventions we have adopted $E < 0$ corresponds to a universe that will recollapse, $E > 0$ corresponds to a universe that will expand forever, and $E = 0$ is the boundary between the two.

To return to (2.9), if we set ν equal to a spatial index the equation is tautological. On the other hand with $\nu = 0$ we have

$$\frac{\partial}{\partial t}[R^3(\Lambda + 8\pi G_N\rho)] = 3R^2\dot{R}[\Lambda - 8\pi G_N P]. \tag{2.19}$$

If Λ is a constant, the usual case, then (2.19) takes the familiar form

$$\frac{\partial}{\partial t}[R^3\rho] = -3R^2\dot{R}P, \tag{2.20}$$

which can also be written as

$$\frac{\partial}{\partial t}\rho = -\frac{3\dot{R}}{R}[\rho + P]. \tag{2.21}$$

We cannot discuss the solutions of these equations until we know ρ and P in various regimes, something we will learn from the kinetic theory. But there is a regime, the de Sitter inflation described by (1.30), where we can make an interesting connection between ρ and P. From (2.7) and (2.8), along with (1.30), we derive the equations

$$\frac{\dot{R}^2}{R^2} = \frac{\ddot{R}}{R}, \tag{2.22}$$

and

$$\frac{2k}{R^2} = 8\pi G_N[\rho + P]. \tag{2.23}$$

In the conventional inflationary picture $k = 0$, so that during inflation

$$\rho = -P, \tag{2.24}$$

and, using (2.21), both ρ and P are constants.

We now turn to the kinetic theory.

3

Tools: kinetic theory

The fundamental object in the kinetic theory description of a fluid of particles in the expanding universe is the density function in phase space of a given species, $f(p, t)$, where $p = |\mathbf{p}|$. Since we are restricting ourselves to Robertson–Walker metrics, f is not a function of either $\mathbf{p}$ or $\mathbf{r}$. Otherwise the space would fail to be homogeneous and isotropic. To maintain the interpretation of f as a density function it must be both real and positive definite. Since the universe evolved from a tiny, extremely dense, singular state one may well ask for what epochs should a phase space description be possible. Presumably it should be possible so long as the de Broglie wavelength of the particles is small compared to the size of the universe. The de Broglie wavelength is given by

$$\lambda\!\!\!^{-} = \hbar/p. \tag{3.1}$$

If we use units in which

$$\hbar = c = k = 1, \tag{3.2}$$

where k is the Boltzmann constant,

$$k = 8.617 \times 10^{-11}\ \mathrm{MeV/K},$$

then if the universe is at a temperature T

$$\lambda\!\!\!^{-} \simeq 1/T. \tag{3.3}$$

To express the lifetime of the universe it is convenient to introduce the so-called Planck mass defined so that, in the units of (3.2)†

$$G_N M^2{}_{\mathrm{pl}} = 1. \tag{3.4}$$

Numerically,

$$M_{\mathrm{pl}} = 2.17 \times 10^{-5}\ \mathrm{g} = 1.2 \times 10^{19}\ \mathrm{GeV} = 1.39 \times 10^{32}\ \mathrm{K} \tag{3.5}$$

† In conventional units $G_N M_{\mathrm{pl}}^2/\hbar c = 1$.

while

$$1/M_{\text{pl}} = 0.55 \times 10^{-43}\ \text{s} = 1.65 \times 10^{-33}\ \text{cm}. \tag{3.6}$$

More generally, for future reference,

$$1\ \text{GeV}^{-1} = 0.197 \times 10^{-13}\ \text{cm} = 0.658 \times 10^{-24}\ \text{s}, \tag{3.7}$$

while

$$1\ \text{GeV} = 1.160 \times 10^{13}\ \text{K}. \tag{3.8}$$

For purposes of our argument we shall make use of a result to be derived later, namely, during the epoch when the energy density of the universe is dominated by massless particles its lifetime is given approximately by

$$t \simeq M_{\text{pl}}/T^2. \tag{3.9}$$

In our units this is also its approximate size.† Thus

$$\lambda\!\!\!^{-}/t \simeq T/M_{\text{pl}}, \tag{3.10}$$

so we expect to be able to use the phase space description so long as

$$T < M_{\text{pl}}.$$

We shall begin our kinetic theory discussion by making the simplifying assumption that $k = 0$. As far as we know, the curvature term does not play a significant role in the evolution of the universe. Leaving it out greatly simplifies the formulae and, if necessary, it can always be reintroduced. Consider a spacelike surface σ. The local normal, $\mathrm{d}S_\mu$, to this surface is defined in such a way that for t = constant, spacelike surfaces,

$$\mathrm{d}S_\mu = (R^3(t)\,\mathrm{d}x_1\,\mathrm{d}x_2\,\mathrm{d}x_3, 0, 0, 0). \tag{3.11}$$

We define $f(x, p)$, for a given species in the fluid, in such a way that the number of world-lines that penetrate $\mathrm{d}\sigma$ is given by

$$\mathrm{d}N = -f(x, p)p^\mu\,\mathrm{d}S_\mu 2\delta^{(+)}(p^2 + m^2)\,\mathrm{d}^4pR^3. \tag{3.12}$$

We have, momentarily, retained the spatial dependence in the f so we can see how the isotropy of space is used to simplify the results. To clarify the meaning of the $\delta^{(+)}$ function we note that

$$p^2 = g_{\mu\nu}p^\mu p^\nu = g^{\mu\nu}p_\mu p_\nu = R^2(t)\mathbf{p}^2 - p_0^2, \tag{3.13}$$

† This "size" is that of the visible horizon.

so that on the mass-shell,

$$R^2(t)\mathbf{p}^2 - p_0^2 = -m^2. \tag{3.14}$$

Thus we can write

$$\delta^{(+)}(p^2 + m^2) = \frac{\delta[p_0 - (m^2 + R^2\mathbf{p}^2)^{\frac{1}{2}}]}{2p_0}. \tag{3.15}$$

This is what is meant by the $\delta^{(+)}$ function occurring in (3.12).

We will be interested in the variation of $f(x, p)$ along a world-line characterized by an affine parameter λ. Thus,

$$\frac{\mathrm{d}}{\mathrm{d}\lambda} f(x(\lambda), p(\lambda)) = \frac{\partial f}{\partial x^\mu}\frac{\mathrm{d}x^\mu}{\partial \lambda} + \frac{\partial f}{\partial p^\mu}\frac{\mathrm{d}p^\mu}{\mathrm{d}\lambda}. \tag{3.16}$$

If we assume that between collisions particles are interacted upon only gravitationally,[†] they then obey the Einstein equation

$$\frac{\mathrm{d}p^\mu}{\mathrm{d}\lambda} = -\Gamma^\mu_{\alpha\beta}p^\alpha p^\beta, \tag{3.17}$$

where

$$p^\mu = \frac{\mathrm{d}x^\mu}{\mathrm{d}\lambda}. \tag{3.18}$$

Hence, using (1.16),

$$\frac{\mathrm{d}p^0}{\mathrm{d}\lambda} = -p^i p^j \Gamma^0{}_{ij} = -\mathbf{p}^2 R\dot{R} \tag{3.19}$$

and

$$\frac{\mathrm{d}p^i}{\mathrm{d}\lambda} = -2\Gamma^i_{0j}p^0 p^j - \Gamma^i_{jk}p^j p^k = -2\frac{\dot{R}}{R}p^0 p^i. \tag{3.20}$$

The overdot here refers to the universal time on a given t = constant, spacelike surface. In (3.20) we have used the vanishing of Γ^i_{jk}. Thus (3.16) can be written,

$$\frac{\mathrm{d}f}{\mathrm{d}\lambda} = p^0\frac{\partial f}{\partial t} + \mathbf{p}\cdot\nabla f - 2\frac{\dot{R}}{R}p^0 p^i\frac{\partial}{\partial p^i}f - \mathbf{p}^2 R\dot{R}\frac{\partial}{\partial p^0}f. \tag{3.21}$$

[†] If there were an electromagnetic background field, $F_{\mu\nu}$, there would be an additional term of the form $F^\mu_\nu p^\nu$ in (3.17).

At this point we shall drop the ∇f term in (3.21) on the grounds of the isotropy of the Robertson–Walker metric. We shall now perform the mass-shell integral

$$\int \frac{\mathrm{d}f}{\mathrm{d}\lambda}\frac{1}{p^0}\,\delta[p_0 - (\mathbf{p}^2R^2 + m^2)^{\frac{1}{2}}]\,\mathrm{d}p^0.$$

For the moment we shall adopt the notation

$$\hat{f}(p,t) = \int \delta[p_0 - (\mathbf{p}^2R^2 + m^2)^{\frac{1}{2}}]f(\mathbf{p},p_0,t)\,\mathrm{d}p_0. \tag{3.22}$$

Since the Boltzmann equation we shall derive will be in terms of f, once we have derived the equation we can drop this extra bit of notation. Upon integration we find

$$\begin{aligned}\int \frac{\mathrm{d}f}{\mathrm{d}\lambda}\frac{1}{p_0}\,\delta[p_0 - (\mathbf{p}^2R^2 + m^2)^{\frac{1}{2}}]\,\mathrm{d}p_0 \qquad\qquad & (3.23)\\ = \int \frac{\partial f}{\partial t}\,\delta[p_0 - (\mathbf{p}^2R^2 + m^2)^{\frac{1}{2}}]\,\mathrm{d}p_0 &\\ -2\frac{\dot{R}}{R}p^i\frac{\partial}{\partial p^i}\hat{f} + \frac{p^2}{p_0}R\dot{R}\left.\frac{\partial f}{\partial p^0}\right|_{p_0=(\mathbf{p}^2R^2+m^2)^{1/2}}. &\end{aligned}$$

We may now use the identity

$$\begin{aligned}\frac{\partial \hat{f}}{\partial t} &= \int \frac{\partial f}{\partial t}\,\delta[p_0 - (\mathbf{p}^2R^2 + m^2)^{\frac{1}{2}}]\,\mathrm{d}p_0 + \int f\frac{\partial}{\partial t}\,\delta[p_0 - (\mathbf{p}^2R^2 + m^2)^{\frac{1}{2}}]\,\mathrm{d}p_0\\ &= \int \frac{\partial f}{\partial t}\,\delta[p_0 - (\mathbf{p}^2R^2 + m^2)^{\frac{1}{2}}]\,\mathrm{d}p_0 + \left.\frac{\partial f}{\partial p_0}\frac{p^2}{p_0}\right|_{p_0=(\mathbf{p}^2R^2+m^2)^{1/2}} \times R\dot{R}\end{aligned} \tag{3.24}$$

Thus

$$\int \frac{\mathrm{d}f}{\mathrm{d}\lambda}\frac{1}{p^0}\,\delta[p_0 - (\mathbf{p}^2R^2 + m^2)^{\frac{1}{2}}]\,\mathrm{d}p_0 = \frac{\partial \hat{f}}{\partial t} - 2\frac{\dot{R}}{R}p^i\frac{\partial \hat{f}}{\partial p^i} \equiv L(\hat{f}). \tag{3.25}$$

This equation serves to define the Liouville operator L appropriate to the flat-space Robertson–Walker cosmology. From now on we will drop the "hat" over the f and write the equations without it. Before continuing, we note that because of the isotropy

$$f = f(|\mathbf{p}|, p_0, R(t)).$$

Hence defining, as before,

$$p = |\mathbf{p}|, \tag{3.26}$$

$$L(f) = \frac{\partial f}{\partial t} - 2\frac{\dot{R}}{R}p\frac{\partial f}{\partial p}.$$

The *collisionless* Boltzmann equation, then, is the statement that

$$L(f) = 0; \tag{3.27}$$

i.e., that there is no net variation of f on the mass-shell along an affine trajectory.

The most general solution to this equation is *any* function of R^2p. In particular the exponential $\exp(-pR^2/T_0)$ is a solution. We can write this as $\exp(-pR/T_0/R)$. Since any function of this quantity is a solution, we have the theorem that a collisionless gas of *massless* particles – Fermi–Dirac or Bose–Einstein – once having obtained an equilibrium distribution can expand freely preserving this form of the distribution provided that $T \sim 1/R$. As we shall see, a gas of *massive* particles has entirely different properties.

The quantity f, the Boltzmann function, is determined by the microscopic behavior of the system – the interparticle collisions. In this work it is convenient to introduce the "local momentum," $\bar{\mathbf{p}} = R\mathbf{p}$. If we regard f as a function of $|\bar{\mathbf{p}}|$ and $\bar{\mathbf{p}}_0 = (\bar{\mathbf{p}}^2 + m^2)^{\frac{1}{2}}$ we can repeat the steps that led to (3.25). From the relation

$$\frac{d\bar{p}^i}{d\lambda} = R\frac{dp^i}{d\lambda} + p^i\frac{dR}{d\lambda}, \tag{3.28}$$

combined with the zeroth component of (3.18), we have

$$\frac{d\bar{p}^i}{d\lambda} = -\frac{\dot{R}}{R}\bar{p}^0\bar{p}^i. \tag{3.29}$$

Hence, in terms of the local momentum,

$$L(f) = \frac{\partial f}{\partial t} - \frac{\dot{R}}{R}\bar{p}_i\frac{\partial}{\partial \bar{p}_i}f. \tag{3.30}$$

From now on we will use only the local momentum and, for convenience of writing, we will drop the bars over the momentum symbols. We shall now introduce some of the macroscopic quantities that can be defined in terms of f.

The particle current density N^μ is defined as

$$N^\mu = \int f \frac{p^\mu}{p^0} \frac{d^3p}{(2\pi)^3}. \tag{3.31}$$

Because of the isotropy the only nonvanishing component of N^μ is N^0, the particle number density n, i.e.,

$$N^0 \equiv n = \int f \frac{d^3p}{(2\pi)^3}. \tag{3.32}$$

But the covariant divergence of a vector field A^μ is given by,

$$A^\mu{}_{;\mu} = \frac{1}{g^{\frac{1}{2}}} \frac{\partial}{\partial x^\mu} (g^{\frac{1}{2}} A^\mu). \tag{3.33}$$

Thus

$$N^\mu{}_{;\mu} = \frac{1}{R^3}\frac{\partial}{\partial t}\left(R^3 \int f \frac{d^3p}{(2\pi)^3}\right) = 3\frac{\dot{R}}{R}\int f \frac{d^3p}{(2\pi)^3} + \int \frac{\partial f}{\partial t}\frac{d^3p}{(2\pi)^3} \tag{3.34}$$

$$= 3\frac{\dot{R}}{R}\int f \frac{d^3p}{(2\pi)^3} + \frac{\dot{R}}{R}\int p_i \frac{\partial}{\partial p_i} f \frac{d^3p}{(2\pi)^3} + \int L(f) \frac{d^3p}{(2\pi)^3}.$$

Integration by parts, dropping the boundary terms, cancels the first and second integrals, leaving

$$N^\mu{}_{;\mu} = \int L(f) \frac{d^3p}{(2\pi)^3} = \frac{1}{R^3}\frac{\partial}{\partial t}(R^3 n). \tag{3.35}$$

This gives the condition for the conservation of the total particle number $R^3 n$.

The conventional definition of the energy–momentum tensor $T^{\mu\nu}$ is given by

$$T^{\mu\nu} = \int f \frac{p^\mu p^\nu}{p_0} \frac{d^3p}{(2\pi)^3}, \tag{3.36}$$

so that

$$T^{00} = \int f p_0 \frac{d^3p}{(2\pi)^3} \equiv \rho.$$

On the other hand,

$$T^{ij} = \int f \frac{p^i p^j}{p_0} \frac{d^3p}{(2\pi)^3}. \tag{3.37}$$

Since the f, in this isotropic universe, is not a function of direction we can integrate over direction and find,

$$T^{ij} = \tfrac{1}{3} g^{ij} \int \frac{fp^2}{p_0} \frac{\mathrm{d}^3 p}{(2\pi)^3} \equiv g^{ij} P. \tag{3.38}$$

Since f is positive definite this pressure P, which we may call the "kinematic pressure," will also be positive. If the objects, of which f is the momentum distribution, interact then there will be an additional pressure – the "dynamical pressure" – which may have either sign. A good example to keep in mind is the van der Waals gas, for which the equation of state is given by

$$P = \frac{N}{\beta V} \frac{1}{(1 - AN/V)} - B \frac{N^2}{V^2}, \tag{3.39}$$

where the sign of P depends on the relationship between A and B, which depends, in turn, on the potential. In applications of (3.38) to cosmology one must keep in mind the possibility of dynamical effects.

Since $T^{\mu\nu}$ is to function as the source term for the Einstein equations we must have

$$T^{\mu\nu}{}_{;\mu} = 0, \tag{3.40}$$

which, referring to (2.20) and assuming $T^{\mu\nu}$ is given entirely by (3.36), means that

$$\frac{1}{R^3} \frac{\partial}{\partial t} \left(R^3 \int f p_0 \frac{\mathrm{d}^3 p}{(2\pi)^3} \right) = - \frac{\dot{R}}{R} \int f \frac{p^2}{p_0} \frac{\mathrm{d}^3 p}{(2\pi)^3}. \tag{3.41}$$

It is not obvious how this condition will be satisfied for an arbitrary solution, f, to the Boltzmann equation. To make things as general as possible we shall introduce a "collision term," $C(E(p))$, which we shall define simply as the right-hand side of the Boltzmann equation

$$L(f) = C(E(p)). \tag{3.42}$$

We can now use the Boltzmann equation as follows:

$$\begin{aligned} \frac{\partial}{\partial t} \left(R^3 \int f p_0 \frac{\mathrm{d}^3 p}{(2\pi)^3} \right) &= 3R^2 \dot{R} \int f p_0 \frac{\mathrm{d}^3 p}{(2\pi)^3} + R^3 \int \frac{\partial f}{\partial t} p_0 \frac{\mathrm{d}^3 p}{(2\pi)^3} \\ &= 3R^2 \dot{R} \int f p_0 \frac{\mathrm{d}^3 p}{(2\pi)^3} + R^3 \int \left(\frac{\dot{R}}{R} p \frac{\mathrm{d}f}{\mathrm{d}p} + C(E) \right) p_0 \frac{\mathrm{d}^3 p}{(2\pi)^3} \\ &= -R^2 \dot{R} \int f \frac{p^2}{p_0} \frac{\mathrm{d}^3 p}{(2\pi)^3} + R^3 \int C(E) p_0 \frac{\mathrm{d}^3 p}{(2\pi)^3}. \end{aligned} \tag{3.43}$$

Thus (3.41) is satisfied provided that

$$\int C(E)p_0 \frac{d^3p}{(2\pi)^3} = 0. \tag{3.44}$$

Let us examine this condition for a simple, but very instructive, case. We consider a single species suffering binary collisions. In this case, ignoring quantum mechanical statistical factors, $C(E(p))$ takes the form

$$\begin{aligned} C(E(p)) = \frac{1}{E(p)} \iiint & (2\pi)^4 \delta^{(4)}(p + p' - p_1 - p_2) \\ & \times [W(p, p'; p_1, p_2) f(p_1) f(p_2) - W(p_1, p_2; p, p') f(p) f(p')] \\ & \times \frac{d^3p'}{(2\pi)^3 2E(p')} \frac{d^3p_1}{(2\pi)^3 2E(p_1)} \frac{d^3p_2}{(2\pi)^3 2E(p_2)}, \end{aligned} \tag{3.45}$$

where $W(p, p'; p_1, p_2)$ is the dimensionless collision probability. To save writing, we introduce the notation

$$dP = \frac{d^3p}{(2\pi)^3 2E(p)}. \tag{3.46}$$

To complete the argument we invoke the unitarity of the S-matrix which, in terms of the Ws means that

$$\begin{aligned} & \iint \delta^{(4)}(p + p' - p_1 - p_2) W(p, p'; p_1, p_2) \, dP_1 P_2 \\ & \quad = \iint \delta^{(4)}(p + p' - p_1 - p_2) W(p_1, p_2; p, p') \, dP_1 P_2. \end{aligned} \tag{3.47}$$

This is a weaker statement than the equality

$$W(p_1, p_2; p, p') = W(p, p'; p_1, p_2), \tag{3.48}$$

which follows from a combination of time reversal and parity symmetry.

From (3.47) it follows that (3.45) can be written as

$$\begin{aligned} C(E(p)) = \frac{1}{E(p)} \iiint & (2\pi)^4 \delta^{(4)}(p + p' - p_1 - p_2) \\ & \times W(p, p'; p_1, p_2)[f(p_1) f(p_2) - f(p) f(p')] \, dP' \, dP_1 \, dP_2. \end{aligned} \tag{3.49}$$

[It is clear, by changing variables, that

$$\iiiint \delta^{(4)}(p + p' - p_1 - p_2) f(p_1) f(p_2) \, dP \, dP' \, dP_1 \, dP_2 \tag{3.50}$$

$$= \iiiint \delta^{(4)}(p + p' - p_1 - p_2) f(p) f(p') \, dP \, dP' \, dP_1 \, dP_2 .]$$

Before completing the argument that will lead to (3.44), it is clear, using (3.47) that

$$\int C(E(p)) \, d^3 p = 0. \tag{3.51}$$

This condition guarantees the conservation of number current. To deal with (3.44) consider

$$\int C(E(p)) E(p) \frac{d^3 p}{2(2\pi)^3} = \iiiint (2\pi)^4 \delta^{(4)}(p + p' - p_1 - p_2) W(p, p'; p_1, p_2)$$
$$\times E(p) [f(p_1) f(p_2) - f(p) f(p')] \, dP \, dP' \, dP_1 \, dP_2 . \tag{3.52}$$

Since the two initial and two final particles are identical we shall assume that W is invariant under exchanges of p and p' as well as exchanges of p_1 and p_2.† Thus we can write

$$\int C(E(p)) E(p) \frac{d^3 p}{2(2\pi)^3} \tag{3.53}$$
$$= \tfrac{1}{2} \iiiint (2\pi)^4 \delta^{(4)}(p + p' - p_1 - p_2) W(p, p'; p_1, p_2)$$
$$\times [E(p) + E(p')] [f(p_1) f(p_2) - f(p) f(p')] \, dP \, dP' \, dP_1 \, dP_2 .$$

† This symmetry is contained in the assumption of Lorentz invariance of the W. In this case

$$W = W(s, t, u),$$

where

$$s = (p_1 + p_2)^2 \quad t = (p_1 - p)^2 \quad u = (p_1 - p')^2$$

and the exchanges described in the text correspond to the tautologies

$$s \to s \quad t \to t \quad u \to u.$$

Now, substituting p for p_1 and p' for p_2, we have,

$$\int C(E(p))E(p)\frac{d^3p}{2(2\pi)^3} = \tfrac{1}{2}\iiiint (2\pi)^4\delta^{(4)}(p + p' - p_1 - p_2) \tag{3.54}$$
$$\times W(p_1, p_2; p, p')[E(p_1) + E(p_2)]$$
$$\times [f(p)f(p') - f(p_1)f(p_2)]\,dP\,dP'\,dP_1\,dP_2.$$

If we now use the conservation of energy, and (3.47), the result follows. In more general cases the conservation of $T^{\mu\nu}$ will involve cancellations among tensors associated with different components of the gas. We turn next to the entropy.

With the quantum mechanical effects included, the entropy density current is defined in terms of f as

$$S^\mu = -\int [f\ln f \mp (1 \pm f)\ln(1 \pm f)]\frac{p^\mu}{p^0}\frac{d^3p}{(2\pi)^3}, \tag{3.55}$$

where the upper and lower signs refer to Bose–Einstein and Fermi–Dirac statistics, respectively. To illustrate general principles we go to the classical limit so that

$$S^\mu = -\int (f\ln f - f)\frac{p^\mu}{p^0}\frac{d^3p}{(2\pi)^3}. \tag{3.56}$$

We are going to take the covariant divergence of S^μ. In this enterprise the second term in (3.56) is irrelevant since it is the conserved number current density. Thus,

$$S^\mu{}_{;\mu} = -\frac{1}{R^3}\frac{\partial}{\partial t}\left(R^3\int f\ln f\frac{d^3p}{(2\pi)^3}\right) \tag{3.57}$$
$$= -3\frac{\dot{R}}{R}\int f\ln f\frac{d^3p}{(2\pi)^3} - \int \dot{f}(\ln f + 1)\frac{d^3p}{(2\pi)^3}$$
$$= -\int C(E)\ln f\frac{d^3p}{(2\pi)^3},$$

where we have used the Boltzmann equation and (3.51). We can now use unitarity to write,

$$S^\mu{}_{;\mu} = -\int C(E)\ln f\frac{d^3p}{(2\pi)^3} \tag{3.58}$$
$$= -\iiiint (2\pi)^4\delta^{(4)}(p + p' - p_1 - p_2)W(p, p'; p_1, p_2)$$
$$\times \ln f(p)[f(p_1)f(p_2) - f(p)f(p')]^2\,dP\,dP'\,dP_1\,dP_2.$$

We can now exploit the various symmetries of W to write,

$$S^{\mu}{}_{;\mu} = -\tfrac{1}{4}\iiiint (2\pi)^4\delta^{(4)}(p + p' - p_1 - p_2) \times W(p, p'; p_1, p_2)\ln\left(\frac{f(p)f(p')}{f(p_1)f(p_2)}\right) \times [f(p_1)f(p_2) - f(p)f(p')]2\,\mathrm{d}P\,\mathrm{d}P'\,\mathrm{d}P_1\,\mathrm{d}P_2. \tag{3.59}$$

Since the fs are positive definite we can use the inequality

$$(x - y)\ln(y/x) \leqslant 0 \tag{3.60}$$

to conclude that

$$S^{\mu}{}_{;\mu} \geqslant 0. \tag{3.61}$$

Both the result and the techniques of this argument are familiar. What is surprising is what results if one attempts to apply the equilibrium condition,

$$S^{\mu}{}_{;\mu} = 0. \tag{3.62}$$

We first apply it using (3.59) to conclude that the putative equilibrium distribution f_{eq} satisfies,

$$\ln f_{\mathrm{eq}}(p) + \ln f_{\mathrm{eq}}(p') = \ln f_{\mathrm{eq}}(p_1) + \ln f_{\mathrm{eq}}(p_2). \tag{3.63}$$

Using the δ-function energy–momentum condition and the scalar nature of f, which is implicit in its definition via (3.12), we can conclude that

$$\ln f_{\mathrm{eq}}(p) = \alpha(t) + \beta^{\mu}(t)p_{\mu}, \tag{3.64}$$

where α is a scalar function of the universal time t and $\beta^{\mu}(t)$ is a *timelike* four-vector. This last is necessary because f_{eq} must be bounded as $\mathbf{p}$ goes to infinity. If β^{μ} were spacelike f_{eq} would have a term like $\mathrm{e}^{\mathbf{p}\cdot\boldsymbol{\beta}}$ and its moments such as $T^{\mu\nu}$ would be ill-defined. Thus

$$f_{\mathrm{eq}}(p) = \mathrm{e}^{\alpha(t) + \beta^{\mu}(t)p_{\mu}}. \tag{3.65}$$

But from (3.57), if (3.62) is to be satisfied we must have

$$L(f_{\mathrm{eq}}) = 0. \tag{3.66}$$

It can be shown on very general grounds – see Appendix A – that, except for special cases to be discussed, (3.66) *cannot* be satisfied in a Robertson–Walker metric. The essence of the general argument, detailed in the appendix, is that the β^{μ} defined by (3.64) must be a spatially constant Killing vector if (3.66) is to be satisfied. Since, as we have shown (recall (1.40)) that in this metric there is no timelike spatially constant Killing vector, the conclusion follows. However, it is more intuitive to study concretely the equation

$$L(f_{eq}) = \frac{\partial f_{eq}}{\partial t} - \frac{\dot{R}}{R} p \frac{\partial}{\partial p} f_{eq} = 0. \tag{3.67}$$

We may, invoking the timelike character of β^μ and the isotropy, write

$$f_{eq}(p, t) = e^{\alpha(t) - \beta(t)E}, \tag{3.68}$$

where $\beta(t)$ is interpreted as the inverse temperature. From (3.67) we find

$$\frac{\dot{\alpha}}{\dot{\beta}} = E(p) - \frac{\dot{R}}{R}\frac{\beta}{\dot{\beta}}\frac{p^2}{E}. \tag{3.69}$$

In general, there are no $\alpha(t)$ or $\beta(t)$ which solve this equation. This is to be expected from the theorem. But there are two limiting cases where there is a solution, namely, $m \to 0$ and $m \to \infty$. In the first case (3.69) becomes

$$\frac{\dot{\alpha}}{\dot{\beta}} = p\left(1 - \frac{\dot{R}}{R}\frac{\beta}{\dot{\beta}}\right). \tag{3.70}$$

This equation has the solution

$$\alpha = 0 \tag{3.71}$$

$$\beta = \text{constant} \times R(t).$$

Hence we recover the result following (3.27), that a gas of massless particles once in equilibrium can continue to expand adiabatically in an equilibrium distribution with $T \sim 1/R$.

In the limit $m \to \infty$ (3.69) becomes

$$\frac{\dot{\alpha}}{\dot{\beta}} - m = \frac{p^2}{2m} - \frac{\dot{R}}{R}\frac{\beta}{\dot{\beta}}\frac{p^2}{m}. \tag{3.72}$$

This can be satisfied with the conditions

$$\alpha - m\beta = \text{constant} \tag{3.73}$$

and

$$\frac{\dot{R}}{R}\frac{\beta}{\dot{\beta}} = \tfrac{1}{2}. \tag{3.74}$$

This means that for the nonrelativistic equilibrium distribution

$$\beta = \text{constant} \times R^2. \tag{3.75}$$

Thus, nonrelativistically, using (3.73) and (3.75)

$$\exp(\alpha - E\beta) = \text{constant} \times \exp\left(\frac{-p^2}{2m}\frac{1}{t_0}\frac{R^2(t)}{R^2(t_0)}\right) \tag{3.76}$$

is an equilibrium distribution. This is also consistent with the discussion following (3.26), allowing for the fact that here we are working in terms of the local momentum.

The fact that, in general, there is no equilibrium distribution for a Robertson–Walker gas has provoked a good deal of discussion in the literature. (See, for example, Schücking and Spiegel, (1970)). For our purposes the question it raises is under what circumstances can the equilibrium distribution be used approximately in cosmological calculations, and how are departures from this distribution to be calculated. The study of these matters will be our main preoccupation in what follows.

4
The canonical example

In this chapter we shall study in detail a one-component system characterized by the Boltzmann equation†

$$\frac{\partial f(p,t)}{\partial t} - \frac{\dot{R}}{R} p \frac{\partial}{\partial p} f(p,t) \tag{4.1}$$
$$= \frac{1}{E(p)} \iiint (2\pi)^4 \delta^{(4)}(p + p' - p_1 - p_2)$$
$$\times W(p, p'; p_1, p_2)[f(p_1)f(p_2) - f(p)f(p')]$$
$$\times \mathrm{d}P' \, \mathrm{d}P_1 \, \mathrm{d}P_2 .$$

We have seen that, in general, there is no function f which makes both sides of this equation vanish. But we can try a solution of the form

$$f(p,t) = f_0(p,t)[1 + \phi(p,t)], \tag{4.2}$$

where

$$f_0(p,t) = \mathrm{e}^{\alpha(t) - \beta(t)E(p)}. \tag{4.3}$$

We want to investigate circumstances of approximate equilibrium in which‡

$$|\phi(p,t)| < 1.$$

The conservation of energy yields the identity

$$f_0(p)f_0(p') = f_0(p_1)f_0(p_2). \tag{4.4}$$

† The work in this section was done in collaboration with L. S. Brown. Related, albeit somewhat more formal, work was done previously by Israel and Vardalas (1970).

‡ For some values of p, ϕ may be negative, provided that f is positive definite for all values of p.

Using this identity and substituting (4.2) we have

$$[f(p_1)f(p_2) - f(p)f(p')] \\ = f_0(p)f_0(p')[\phi(p_1) + \phi(p_2) - \phi(p) - \phi(p')] + O(\phi^2). \tag{4.5}$$

If we solve the Boltzmann equation with this approximate kernel there is an ambiguity in the definition of ϕ. To any ϕ one can associate a ϕ' defined by

$$\phi' = \phi + a + bE(p), \tag{4.6}$$

where a and b are arbitrary functions of time but not of momentum. Because of the conservation of energy this ϕ' produces, to this order, the same kernel – (4.5) – as does ϕ. This ambiguity is resolved by invoking two "gauge conditions"; namely,

$$\int f_0 \phi \, \mathrm{d}^3 p = \int E f_0 \phi \mathrm{d}^3 p = 0. \tag{4.7}$$

Physically this means that the number density n is given by

$$n = \int f_0 \frac{\mathrm{d}^3 p}{(2\pi)^3}, \tag{4.8}$$

and the energy density is given by

$$\rho = \int E f_0 \frac{\mathrm{d}^3 p}{(2\pi)^3}. \tag{4.9}$$

Both of these integrals can actually be evaluated in terms of Bessel functions but the result is not especially enlightening.

For the pressure defined as

$$P = \tfrac{1}{3} \int f \frac{p^2}{E} \frac{\mathrm{d}^3 p}{(2\pi)^3} \tag{4.10}$$

there are no constraint conditions. It is just this, as we shall now see, that leads to the curious relativistic phenomenon known as "bulk viscosity," the generation of entropy by the expanding gas. We can, using (5.2), write

$$P = P_0 + P_1, \tag{4.11}$$

where

$$P_0 = \tfrac{1}{3} \int \frac{p^2}{E} f_0 \frac{\mathrm{d}^3 p}{(2\pi)^3} \tag{4.12a}$$

and

$$P_1 = \frac{1}{3}\int \frac{p^2}{E} f_0 \phi \frac{d^3p}{(2\pi)^3}. \tag{4.12b}$$

To see that this gas generates entropy we use (4.59) with the distribution given by (4.2). Thus,

$$\begin{aligned} S^{\mu}{}_{;\mu} = -\tfrac{1}{4} \iiiint &(2\pi)^4 \delta^{(4)}(p + p' - p_1 - p_2) \\ &\times W(p,p';p_1,p_2) \ln\left[\frac{[1+\phi(p)][1+\phi(p')]}{[1+\phi(p_1)][1+\phi(p_2)]}\right] f_0(p_1) f_0(p_2) \\ &\times [\phi(p_1)+\phi(p_2)-\phi(p)-\phi(p')] 2\,dP\,dP'\,dP_1\,dP_2 \\ \simeq \tfrac{1}{4} \iiiint &(2\pi)^4 \delta^{(4)}(p + p' - p_1 - p_2) W(p,p';p_1,p_2) \\ &\times f_0(p_1) f_0(p_2) [\phi(p_1)+\phi(p_2)-\phi(p)-\phi(p')]^2 \\ &\times 2\,dP\,dP'\,dP_1\,dP_2. \end{aligned} \tag{4.13}$$

The fact that entropy generation is a second-order effect in ϕ suggests an iterative approach to the solution of the Boltzmann equation; namely, we employ a first iterate using entropy conservation as a condition. The self-consistency of this procedure will require $|\phi| < 1$ and this, as we shall see, will require that

$$\frac{\dot{R}/R}{\Gamma} < 1,$$

where Γ is the reaction rate of the process. We shall therefore neglect terms of order $\phi\dot{R}/R$, as compared to $\Gamma\phi$. This will be relevant when we evaluate $L(f)$. Before doing this it is instructive to use (3.56) to evaluate the entropy density, s, which we take to be S^0 for the distribution given by (4.2). Thus

$$\begin{aligned} s &= -\int f(\ln f - f) \frac{d^3p}{(2\pi)^3} \\ &= \frac{1}{T}\int f_0(1+\phi)(E - \alpha T) \frac{d^3p}{(2\pi)^3} \\ &\quad + \int f_0(1+\phi)[1 - \ln(1+\phi)] \frac{d^3p}{(2\pi)^3} \\ &= \frac{\rho}{T} - n\frac{\mu}{T} + \int f_0 \frac{d^3p}{(2\pi)^3} + O(\phi^2). \end{aligned} \tag{4.14}$$

To conform to custom we have defined the chemical potential μ by the relation

$$\alpha T \equiv \mu. \tag{4.15}$$

We may now invoke the identity,

$$P_0 = \tfrac{1}{3}\int e^{(\alpha-\beta E)}\frac{p^2}{E}\frac{d^3p}{(2\pi)^3} = -\frac{1}{3\beta}\int p\frac{d}{dp}e^{(\alpha-\beta E)}\frac{d^3p}{(2\pi)^3} \tag{4.16}$$
$$= \frac{1}{\beta}\int e^{(\alpha-\beta E)}\frac{d^3p}{(2\pi)^3} = \frac{n_0}{\beta},$$

where we have integrated by parts, dropping the boundary terms. Thus we have

$$s = \frac{\rho + P_0 - \mu n}{T} + O(\phi^2). \tag{4.17}$$

Hence, to order ϕ, the expansion is adiabatic and the thermodynamic functions take their familiar forms.† We now turn to the evaluation of the Liouville operator acting on f, to order ϕ. Since we drop terms of order $\phi\dot{R}/R$ this comes down to evaluating $L(f_0)$. We write f_0 in the form

$$f_0 = Ae^{-\beta E}, \tag{4.18}$$

where A and β are functions of t. Thus

$$L(f_0) = f_0\left[\frac{\dot{A}}{A} - \dot{\beta}E + \frac{\dot{R}}{R}\frac{p^2}{E}\beta\right]. \tag{4.19}$$

This is to be equated to the collision term taken to lowest nontrivial order in ϕ, namely,

$$L(f_0) = \frac{1}{E(p)}\iiint(2\pi)^4\delta^{(4)}(p + p' - p_1 - p_2) \tag{4.20}$$
$$\times W(p, p'; p_1, p_2)f_0(p, t)f_0(p', t)[\phi(p_1) + \phi(p_2)$$
$$- \phi(p) - \phi(p')]\,dP'\,dP_1\,dP_2.$$

We shall now invoke particle number conservation, which, because of (4.7), takes the form

$$\frac{\partial}{\partial t}\left(R^3A\int e^{-\beta E}\frac{d^3p}{(2\pi)^3}\right) = 0. \tag{4.21}$$

† See Appendix B.

This leads to the condition that

$$\frac{\dot{A}}{A} = -3\frac{\dot{R}}{R} + \dot{\beta}\frac{U}{N}, \tag{4.22}$$

where

$$U = R^3\rho \tag{4.23}$$

and

$$N = R^3 n. \tag{4.24}$$

The entropy conservation condition implies that

$$\frac{\partial}{\partial t}\left(R^3 \int f_0 \ln f_0 \frac{\mathrm{d}^3 p}{(2\pi)^3}\right) = 0 \tag{4.25}$$

$$= N\frac{\dot{A}}{A} - \frac{\partial}{\partial t}(\beta U).$$

The two conditions together imply the relation

$$\frac{\partial}{\partial t}\left(\beta\frac{U}{N}\right) = -3\frac{\dot{R}}{R} + \dot{\beta}\frac{U}{N}. \tag{4.26}$$

This condition instructs us as to how β and R are related in various regimes so as to produce the quasi-adiabatic expansion. It is useful to evaluate it in two regimes: high T/m and low T/m. To this end we need n and ρ in these limits.

For $T/m \gg 1$,

$$n = \frac{A}{2\pi^2}\int_0^\infty p^2 \mathrm{e}^{-\beta(p+m^2/2p)}\,\mathrm{d}p \tag{4.27}$$

$$= \frac{A}{2\pi^2}\left(\frac{2}{\beta^3} - \frac{m^2}{2\beta}\right),$$

while

$$\rho = \frac{A}{2\pi^2}\int_0^\infty p^2\left(p + \frac{m^2}{2p}\right)\left(1 - \frac{\beta m^2}{2p}\right)\mathrm{e}^{-\beta p}\,\mathrm{d}p \tag{4.28}$$

$$= 3\frac{n}{\beta} + \tfrac{1}{2}\beta m^2 n.$$

Note that we can write (4.25) in the form

$$\frac{\dot{A}}{A} = \frac{\partial}{\partial t}\left(\frac{\beta\rho}{n}\right), \tag{4.29}$$

so that in the strictly massless limit we recover the fact that $\rho/n \sim T$, since A is a constant in this limit.[†]

For $T/m \ll 1$

$$n = \frac{A}{2\pi^2}\int_0^\infty p^2 e^{-\beta(m + p^2/2m - p^4/8m^2)}\,dp \tag{4.30}$$

$$= \frac{A}{4\pi^2} e^{-\beta m} \frac{m(2\pi m)^{\frac{1}{2}}}{\beta^{3/2}}\left[1 + \frac{15}{8}\frac{1}{\beta m}\right],$$

and

$$\rho = \frac{A}{2\pi^2}\int_0^\infty \left(m + \frac{p^2}{2m}\right) e^{-\beta(m + p2/2m)} p^2\,dp \tag{4.31}$$

$$= mn + \tfrac{3}{2}\frac{n}{\beta}.$$

We can now solve (4.26) in the form

$$\beta\frac{\dot{U}}{N} = -3\frac{\dot{R}}{R}. \tag{4.32}$$

For $T/m \gg 1$, noting that, since N is constant,

$$\left(\frac{\dot{U}}{N}\right) = \left(\frac{\dot{\rho}}{n}\right) \tag{4.33}$$

and that

$$\frac{d}{dt}\ln\left(\frac{\beta}{R}\right) = \frac{\dot{\beta}}{\beta} - \frac{\dot{R}}{R}, \tag{4.34}$$

we have, where c is a constant,

$$\beta/R = c(1 + \tfrac{1}{12}\beta^2 m^2) + O(\beta^4 m^4) \tag{4.35}$$

or

$$TR = c(1 - \tfrac{1}{12}\beta^2 m^2), \tag{4.36}$$

which expresses the fact that the temperature falls more rapidly than in the $m = 0$ limit $T \sim 1/R$. Indeed, where $T/m < 1$, (4.32) becomes simply

$$\frac{\dot{R}}{R} = \frac{1}{2}\frac{\dot{\beta}}{\beta}, \tag{4.37}$$

[†] The fact that A is constant in this limit follows from (4.19) setting the right-hand side equal to zero.

which is the statement that in this limit

$$TR^2 = \text{constant}. \tag{4.38}$$

Hence if we begin with an effectively massless gas for $T/m > 1$, so long as ϕ is small, it will evolve adiabatically into a nonrelativistic gas with $T \sim 1/R^2$. We turn next to the conditions that determine the smallness of ϕ.

To this end, we need to evaluate the expression in parentheses on the right-hand side of (4.19), making use of (4.29). In the relativistic limit, $T > m$, we find, keeping the leading terms

$$L(f_0) = f_0 \frac{\dot{R}}{R} \frac{\beta^2 m^2}{E} \left(E - \frac{1}{\beta} - \beta \frac{E^2}{6} \right). \tag{4.39}$$

In the absence of a particular model we cannot write down the right-hand side of the Boltzmann equation specifically. We shall deal with concrete examples in the sections to come. Our concern here will be with orders of magnitude. We want to estimate ϕ in order to determine the conditions under which the iteration scheme has a chance of converging. First we simplify (4.39) by assuming that in the terms multiplying f_0 we can take $E \simeq 1/\beta$. Thus (4.39) becomes simply,

$$L(f_0) \simeq -f_0 \frac{\dot{R}}{R} \frac{\beta^2 m}{6}. \tag{4.40}$$

The collision term $C(E)$ we estimate as follows:

$$\begin{aligned} C(E) = {} & \frac{1}{E(p)} \iiint (2\pi)^4 \delta^{(4)}(p + p' - p_1 - p_2) \\ & \times W(p, p'; p_1, p_2) f_0(p) f_0(p') [\phi(p_1) + \phi(p_2) \\ & - \phi(p) - \phi(p')] \, \mathrm{d}P' \, \mathrm{d}P_1 \, \mathrm{d}P_2 \\ \simeq {} & n \langle \sigma v \rangle \langle \phi \rangle f_0(p). \end{aligned} \tag{4.41}$$

The last step is done by replacing the ϕ bracket by some average ϕ. Next we approximate the energy factors $E(p)$ and $E(p')$ by $1/\beta$ and carry out the p' integral. Finally we use the expression

$$\begin{aligned} & \iint (2\pi)^2 \delta^{(4)}(p + p' - p_1 - p_2) W(p, p'; p_1, p_2) \, \mathrm{d}P_1 \, \mathrm{d}P_2 \\ & = \langle s\sigma(s, \theta) v \rangle, \end{aligned} \tag{4.42}$$

where s is the square of the total center of mass energy and v is the relative velocity. We replace s approximately by $1/\beta^2$, canceling the energies in

(4.41) and producing the final equation in (4.41). The reaction rate $1/\tau$ is then given by

$$1/\tau = n\langle \sigma v \rangle, \tag{4.43}$$

where τ is the mean time between collisions. Thus, in the $T \gg m$ regime

$$|\phi| \simeq \frac{\tau}{6} \frac{\dot{R}}{R} \beta^2 m^2. \tag{4.44}$$

In the $T < m$ regime we can exploit (4.29), (4.31), and (4.36) to learn that close to equilibrium

$$\frac{\dot{A}}{A} = 2 \frac{\dot{R}}{R} \beta m. \tag{4.45}$$

Thus, expanding to leading order the expression in (4.19), we learn that for $T \ll m$

$$L(f_0) = -f_0 \frac{\dot{R}}{R} \frac{1}{\beta^2 m^2}, \tag{4.46}$$

or, that in this regime

$$|\phi| \simeq \frac{1}{n\langle \sigma v \rangle} \frac{\dot{R}}{R} < 1. \tag{4.47}$$

Most of the treatments in the literature of bulk viscosity have been macroscopic, for example, Weinberg (1971), and it is important to make contact with this work. To this end we need to construct the energy–momentum tensor of a so-called perfect fluid . For this purpose one defines a velocity four-vector U^α such that in the coordinate system expanding with the fluid

$$\mathbf{U} = 0 \tag{4.48}$$

and

$$U^0 = 1. \tag{4.49}$$

Thus

$$U_\alpha U^\alpha = -1. \tag{4.50}$$

A "perfect fluid" is defined to have a $T^{\mu\nu}$ of the form

$$T^{\mu\nu} = P_0 g^{\mu\nu} + (P_0 + \rho) U^\mu U^\nu, \tag{4.51}$$

where, to make contact with the microphysics, P_0 and ρ are defined by (3.37) with f taken to be f_0 so that, to this order, entropy is conserved. To this $T^{\mu\nu}$

one now adds a $\Delta T^{\mu\nu}$ which expresses dissipation. The so-called first-order theory involves only first derivatives of U^μ and T in $\Delta T^{\mu\nu}$. Because of the homogeneity and isotropy of the Robertson–Walker world there are only two scalars one can make which are first order in the derivatives, namely,

$$\frac{1}{T} U^\beta \frac{\partial T}{\partial x^\beta} = \frac{\dot{T}}{T}, \tag{4.52}$$

and

$$U^\alpha{}_{;\alpha} = 3\frac{\dot{R}}{R}. \tag{4.53}$$

To this order we may make use of (4.32), which expresses the combined entropy and particle number conservation. Thus, $\dot{T}/T$ and $\dot{R}/R$ are related by a function of T, such as (4.35); i.e.,

$$\frac{\dot{T}}{T} = \frac{\dot{R}}{R} f(T). \tag{4.54}$$

Hence, up to a function of T, there is only one scalar invariant which we take to be $U^\alpha{}_{;\alpha}$.

Following the discussion preceding (4.9) we demand that

$$\Delta T^{00} = 0. \tag{4.55}$$

We satisfy all constraints if we take

$$\Delta T^{\mu\nu} = -\xi(g^{\mu\nu} + U^\mu U^\nu) U^\alpha{}_{;\alpha}. \tag{4.56}$$

We will shortly show that ξ is a positive definite function of the temperature as a consequence of the second law. It is what is called the "coefficient of bulk viscosity." Thus, in effect, the bulk viscosity is equivalent to the addition of a negative pressure to P_0. Including the bulk viscosity, $T^{\mu\nu}$ can be written as

$$T^{\mu\nu} = \left(P_0 - 3\frac{\dot{R}}{R}\right) g^{\mu\nu} \tag{4.57}$$
$$+ \left(\rho + P_0 - 3\frac{\dot{R}}{R}\xi\right)(g^{\mu\nu} + U^\mu U^\nu).$$

We may now make the contact with the microphysics. From (4.12a) and (4.12b) we can write

$$T^{ij} = g^{ij}\left(\tfrac{1}{3}\int f_0 \frac{p^2 d^3p}{E(p)(2\pi)^3} + \tfrac{1}{3}\int f_0 \phi \frac{p^2}{E(p)} \frac{d^3p}{(2\pi)^3}\right). \tag{4.58}$$

By comparing with (4.57) we may conclude that

$$\xi = -\tfrac{1}{9}\frac{R}{\dot{R}}\int f_0\phi \frac{p^2}{E(p)}\frac{d^3p}{(2\pi)^3}. \tag{4.59}$$

Using the constraint equation (4.7), and the approximate value of $|\phi|$ given by (4.44), we can estimate ξ in the $T \gg m$ regime. Thus

$$\begin{aligned}\xi &\simeq \tfrac{1}{9}\frac{R}{\dot{R}}|\phi|n\frac{m^2\beta^2}{2\beta} \\ &\simeq \frac{n\beta^3\tau m^4}{108}.\end{aligned} \tag{4.60}$$

It is interesting to compare this result to the evaluation of the same quantity given by Weinberg (1971) using the macrophysics. His answer, written in a notation making it easier to compare, is

$$\xi = 4\rho\tau\left[\tfrac{1}{3} - \left[\frac{\partial P}{\partial \rho}\right]_n\right]^2. \tag{4.61}$$

We may use the relations, appropriate to $T \gg m$,

$$P = nT, \tag{4.62}$$

and

$$\rho = 3nT + \tfrac{1}{2}m^2n/T, \tag{4.63}$$

to compute $\partial P/\partial\rho\,|_n$; i.e.,

$$\frac{\partial P}{\partial \rho}\Big|_n = \frac{\partial P/\partial T\,|_n}{\partial\rho/\partial T\,|_n} = \frac{n}{3n - \frac{1}{2}m^2n/T^2} \simeq \tfrac{1}{3}\left[1 + \frac{m^2}{6T^2}\right]. \tag{4.64}$$

Putting terms together, Weinberg finds

$$\xi = \frac{4n\beta^3\tau m^4}{108}, \tag{4.65}$$

which is to be compared to (4.60).

We shall conclude this section by comparing the macroscopic and microscopic versions of the H-theorem and, in particular, by justifying the positivity of ξ. We begin with the macroscopic $T^{\mu\nu}$ written in terms of ρ and P where P includes the bulk viscosity correction. Thus

$$T^{\mu\nu} = Pg^{\mu\nu} + (P+\rho)U^\mu U^\nu. \tag{4.66}$$

This quantity, which is to function as the source term in the Einstein equation, must be covariantly conserved. Since

$$g^{\mu\nu}{}_{;\mu} = 0, \tag{4.67}$$

we must have

$$P_{;\mu}g^{\mu\nu} + (P + \rho)_{;\mu}U^{\mu}U^{\nu} + (P + \rho)(U^{\mu}U^{\nu})_{;\mu} = 0. \tag{4.68}$$

If we multiply by U_ν we have,

$$P_{;\mu}U^{\mu} - (\rho + P)_{;\mu}U^{\mu} + U_\nu(\rho + P)(U^{\mu}U^{\nu})_{;\mu} = 0. \tag{4.69}$$

But from

$$U_\mu U^\mu = -1, \tag{4.70}$$

we have

$$U_\mu U^\mu{}_{;\nu} = 0. \tag{4.71}$$

Thus (4.69) takes the form

$$U^\mu P_{;\mu} - (U^\mu(\rho + P))_{;\mu} = 0. \tag{4.72}$$

The conservation of particle number can be written as

$$(nU^\mu)_{;\mu} = 0, \tag{4.73}$$

so that

$$U^\mu{}_{;\mu} = -\frac{U^\mu}{n} n_{;\mu}. \tag{4.74}$$

Using (4.73) we have the relation

$$[(\rho + P)U^\mu]_{;\mu} = \left(\frac{\rho + P}{n}\right)_{;\mu} U^\mu n. \tag{4.75}$$

Using this we can then write (4.72) as

$$U^\mu\left[P_{;\mu} - n\left(\frac{\rho + P}{n}\right)_{;\mu}\right] = -nU^\mu\left[\left(\frac{\rho}{n}\right)_{;\mu} + P\left(\frac{1}{n}\right)_{;\mu}\right] = 0. \tag{4.76}$$

If we call the specific entropy σ, the entropy per particle, the entropy density current takes the form

$$S^\mu = nU^\mu\sigma. \tag{4.77}$$

The second law written in terms of specific entropy is

$$T\,\mathrm{d}\sigma = \mathrm{d}(\rho/n) + P_0\,\mathrm{d}(1/n). \tag{4.78}$$

Thus using (4.53), (4.57), and (4.74), equation (4.76) is the statement that

$$S^{\mu}{}_{;\mu} = \frac{9}{T}\left(\frac{\dot{R}}{R}\right)^2 \xi. \tag{4.79}$$

Hence the second law requires that $\xi > 0$. Since, using (4.73),

$$S^{\mu}{}_{;\mu} = (nU^{\mu}\sigma)_{;\mu} = n\dot{\sigma}, \tag{4.80}$$

we have an expression for the change in the specific entropy; i.e.,

$$\frac{\dot{\sigma}}{\sigma} = \frac{9}{T}\left(\frac{\dot{R}}{R}\right)^2 \frac{\xi}{n}\frac{1}{\sigma}. \tag{4.81}$$

Using the statistical mechanical evaluation of ξ we have, in the $T \gg m$ regime,

$$\frac{\dot{\sigma}}{\sigma} \simeq \frac{\tau}{12}\left(\frac{\dot{R}}{R}\right)^2 \frac{\beta^4 m^4}{\sigma}. \tag{4.82}$$

This is, as it must be, essentially the same answer we find if we use (4.13) directly. This formula, or its equivalent, has been used by Weinberg (1971), and others, to show that conventional processes in the early universe generate only small amounts of entropy. We now turn our attention to some specific processes.

5

The generalized Lee–Weinberg problem: background

What we shall call the "generalized Lee–Weinberg problem" has the following elements: we imagine a gas of *stable* particles and antiparticles which we call L and $\bar{L}$. As the process begins, we suppose that the particles and antiparticles are kept in tight equilibrium by collisions among themselves or with other particles. Hence, if that were all that was happening, the treatment of the last section would apply. However, and this is the new element, we suppose that the L and $\bar{L}$ can mutually annihilate. For $T > m_L$, the L mass, annihilations are compensated for by creations and hence the equilibrium is maintained. However, as the temperature drops, there will come a temperature that depends on m_l, the mass of the light particles which are the products of the L annihilation, such that the light l particles no longer have enough average energy to reconstitute the L particles. At this critical temperature, equilibrium is no longer maintained. The "generalized Lee–Weinberg problem" is to analyze the statistical mechanics of this process using the relevant Boltzmann equations. Before turning to this we give three examples in which the Lee–Weinberg problem is relevant to cosmology.

(1) The original Lee–Weinberg problem (Lee and Weinberg, 1977) concerned a gas of hypothetical heavy neutral leptons – "neutrinos" – which interact with conventional leptons with the ordinary weak couplings. Some of those heavy leptons would presumably still be with us. Hence they could, in principle, contribute to the present energy density of the universe. If there are too many of them they would dominate the density and could overclose the universe. The problem is then to determine as a function of the masses and coupling constants how many such particles would be around today.

(2) The grand unified models of weak and strong interactions predict the existence of superheavy magnetic monopoles. These objects, theoretically, should have masses the order of the unification temperature – the temperature in the early universe at which the strong, weak, and electromagnetic interactions acquire the same universal strength. This temperature is

supposed to be about 10^{14} GeV. In important papers Zel'dovich and Khlopov (1978) and Preskill (1979) computed how many monopoles would be left over today. They found that this number would be unacceptably large. This was one of the motivations for Guth's introduction of the inflationary universe (Guth, 1981). This is again a system of the Lee–Weinberg type.

(3) This example differs from the first two in some important respects. In the first place, it involves known particles, electrons, photons, neutrinos, and the like, rather than speculative objects like monopoles or heavy leptons. In the second place, the questions we want to ask are different. The process we have in mind is helium production in the early universe. In the standard calculation of helium production (see, for example, Peebles, 1966) the weak interaction rates which determine the neutron–proton ratio as a function of temperature are computed using the equilibrium distributions for the electrons and other leptons. To understand the question that this raises it is useful to recall that in our units

$$1\,\text{MeV} = 1.16 \times 10^{10}\,\text{K}.$$

Now, helium production begins with the formation of deuterium nuclei which have binding energies of 2.225 MeV. Deuteron formation cannot take place in any appreciable amount until the ambient temperature of the photons is less than this binding energy. A calculation shows that the critical temperature is about 10^9 K; substantially lower than the binding energy because otherwise there are enough high energy photons in the tail of the Planck distribution to disintegrate the newly formed deuterons. But the electron mass corresponds to a temperature of this order of magnitude and hence at these temperatures the electrons are passing out of equilibrium. It now becomes a matter of calculation – a calculation we shall do – to determine how large these nonequilibrium effects are.

Before turning to these actual problems it is instructive to consider a model problem which illustrates the novel features of the multicomponent fluid situation. To this end we imagine a Lee–Weinberg problem with the following simplifying elements:

1. We assume that L is identical to $\bar{\text{L}}$ and that l is identical to $\bar{\text{l}}$.
2. We use classical statistics for all the particles.
3. We assume time reversability and parity symmetry.
4. We assume that the only interactions are elastic scatterings of L particles with L particles and l particles with l particles, along with the annihilation reaction $\text{L} + \bar{\text{L}} \rightarrow \text{l} + \bar{\text{l}}$.

We shall call f the distribution functions of the L particles and g the distribution function of the l particles. We may now write down the *coupled* Boltzmann equations for the fs and gs; i.e.,

$$\frac{\partial f}{\partial t} - \frac{\dot{R}}{R} p \frac{\partial f}{\partial p} = C_{\mathrm{E}}^{\mathrm{L}} + C_{\mathrm{I}}^{\mathrm{L}}, \tag{5.1}$$

where

$$C_{\mathrm{E}}^{\mathrm{L}} = \frac{1}{E(p)_{\mathrm{L}}} \iiint (2\pi)^4 \delta^{(4)}(p + p' - p_1 - p_2) \times [W_{\mathrm{E}}^{\mathrm{L}}(p, p'; p_1, p_2) f(p_1) f(p_2) - W_{\mathrm{E}}^{\mathrm{L}}(p_1, p_2; p, p') f(p) f(p')] \times \mathrm{d}P' \,\mathrm{d}P_1 \,\mathrm{d}P_2 \tag{5.2}$$

and

$$C_{\mathrm{I}}^{\mathrm{L}} = \frac{1}{E(p)_{\mathrm{L}}} \iiint (2\pi)^4 \delta^{(4)}(p + p' - q_1 - q_2) \times [W_{\mathrm{I}}(p, p'; q_1, q_2) g(q_1) g(q_2) - W_{\mathrm{I}}(q_1, q_2; p, p') f(p) f(p')] \times \mathrm{d}P' \,\mathrm{d}Q_1 \,\mathrm{d}Q_2. \tag{5.3}$$

For the inelastic terms we have not distinguished between the transition rate for the g and f equations since these, at least in this model, are the same. Using the symmetries of the Ws we can write

$$C_{\mathrm{E}}^{\mathrm{L}} = \frac{1}{E(p)_{\mathrm{L}}} \iiint (2\pi)^4 \delta^{(4)}(p + p' - p_1 - p_2) \times W_{\mathrm{E}}^{\mathrm{L}}(p, p'; p_1, p_2)[f(p_1) f(p_2) - f(p) f(p')] \,\mathrm{d}P' \,\mathrm{d}P_1 \,\mathrm{d}P_2, \tag{5.4}$$

while

$$C_{\mathrm{I}}^{\mathrm{L}} = \frac{1}{E(p)_{\mathrm{L}}} \iiint (2\pi)^4 \delta^{(4)}(p + p' - q_1 - q_2) \times W_{\mathrm{I}}(p, p'; q_1, q_2)[g(q_1) g(q_2) - f(p) f(p')] \,\mathrm{d}P' \,\mathrm{d}Q_1 \,\mathrm{d}Q_2. \tag{5.5}$$

On the other hand the functions g obey the equation

$$\frac{\partial g}{\partial t} - \frac{\dot{R}}{R} q \frac{\partial g}{\partial q} = C_{\mathrm{E}}^{\mathrm{l}} + C_{\mathrm{I}}^{\mathrm{l}}, \tag{5.6}$$

where, exploiting the symmetries of the Ws,

$$C_{\mathrm{E}}^{\mathrm{l}} = \frac{1}{E(q)_{\mathrm{l}}} \iiint (2\pi)^4 \delta^{(4)}(q + q' - q_1 - q_2) \times W_{\mathrm{E}}^{\mathrm{l}}(q, q'; p_1, p_2)[g(q_1) g(q_2) - g(q) g(q')] \,\mathrm{d}Q' \,\mathrm{d}Q_1 \,\mathrm{d}Q_2 \tag{5.7}$$

and

$$C_{\mathrm{I}}^{\mathrm{I}} = \frac{1}{E(q)_{\mathrm{I}}} \iiint (2\pi)^4 \delta^{(4)}(q + q' - p_1 - p_2)$$
$$\times W_{\mathrm{I}}(q, q'; p_1, p_2)[f(p_1)f(p_2) - g(q)g(q')]\, \mathrm{d}Q'\, \mathrm{d}P_1\, \mathrm{d}P_2. \quad (5.8)$$

We may now attempt to repeat the discussion in Chapter 4 which led to the conservation of particle number, $T^{\mu\nu}$, and the H-theorem. We begin with the particle number. In an obvious notation

$$N_{\mathrm{L}}^{\mu} = \int f \frac{p^{\mu}}{p^0} \frac{\mathrm{d}^3 p}{(2\pi)^3}, \quad (5.9)$$

while

$$N_{\mathrm{I}}^{\mu} = \int g \frac{q^{\mu}}{q^0} \frac{\mathrm{d}^3 q}{(2\pi)^3}. \quad (5.10)$$

Following the discussion in Chapter 4 we can conclude that

$$N_{\mathrm{L}}{}^{\mu}{}_{;\mu} = \int (C_{\mathrm{E}}^{\mathrm{L}}(p) + C_{\mathrm{I}}^{\mathrm{L}}(p)) \frac{\mathrm{d}^3 p}{(2\pi)^3}, \quad (5.11)$$

while

$$N_{\mathrm{I}}{}^{\mu}{}_{;\mu} = \int (C_{\mathrm{E}}^{\mathrm{L}}(q) + C_{\mathrm{I}}^{\mathrm{I}}(q)) \frac{\mathrm{d}^3 q}{(2\pi)^3}. \quad (5.12)$$

If we examine the integral on the right-hand sides of (5.2) and (5.3) it becomes clear that neither of the currents is *separately* conserved. This is hardly surprising since particles are being created and annihilated. However, we do have the conservation of *total* particle number; i.e.,

$$N_{\mathrm{L}}{}^{\mu}{}_{;\mu} + N_{\mathrm{I}}{}^{\mu}{}_{;\mu} = 0. \quad (5.13)$$

We can, in a similar way, study the energy-momentum tensors. Thus,

$$T_{\mathrm{L}}{}^{\mu 0}{}_{;\mu} = \int [C_{\mathrm{E}}^{\mathrm{L}}(p) + C_{\mathrm{I}}^{\mathrm{L}}(p)] E(p)_{\mathrm{L}} \frac{\mathrm{d}^3 p}{(2\pi)^3}$$
$$= \tfrac{1}{2} \iiiint [E(p)_{\mathrm{L}} + E(p')_{\mathrm{L}}](2\pi)^4 \delta^{(4)}(p + p' - q_1 - q_2)$$
$$\times W_{\mathrm{I}}(p, p'; q_1, q_2)[g(q_1)g(q_2) - f(p)f(p')]\, \mathrm{d}P\, \mathrm{d}P'\, \mathrm{d}Q_1\, \mathrm{d}Q_2. \quad (5.14)$$

We have taken advantage of the symmetries to cancel the elastic contribution and to write the inelastic contribution as shown. It is clear that while we can evoke the conservation of energy in (5.14) the two terms do not cancel. Likewise we can write,

$$T_{\mathrm{I}}{}^{\mu 0}{}_{;\mu} = \tfrac{1}{2}\iiiint (2\pi)^4\delta^{(4)}(p + p' - q_1 - q_2)[E(q_1)_{\mathrm{I}} + E(q_2)_{\mathrm{I}}]$$
$$\times\, W_{\mathrm{I}}(q_1, q_2; p, p')[f(p)f(p') - g(q_1)g(q_2)]\, \mathrm{d}P\, \mathrm{d}P'\, \mathrm{d}Q_1\, \mathrm{d}Q_2. \tag{5.15}$$

It is now clear, upon examination, that

$$T_{\mathrm{L}}{}^{\mu\nu}{}_{;\mu} + T_{\mathrm{I}}{}^{\mu\nu}{}_{;\mu} = 0. \tag{5.16}$$

Next we consider the entropy. Exploiting the symmetries

$$S_{\mathrm{L}}{}^{\mu}{}_{;\mu} = -\tfrac{1}{4}\iiiint (2\pi)^4\delta^{(4)}(p + p' - p_1 - p_2)$$
$$\times\, W_{\mathrm{E}}^{\mathrm{L}}(p, p'; p_1, p_2) \ln\left[\frac{f(p)f(p')}{f(p_1)f(p_2)}\right]$$
$$\times\, [f(p_1)f(p_2) - f(p)f(p')]\, 2\, \mathrm{d}P\, \mathrm{d}P'\, \mathrm{d}P_1\, \mathrm{d}P_2$$
$$+ \tfrac{1}{2}\iiiint (2\pi)^4\delta^{(4)}(p + p' - q_1 - q_2) W_{\mathrm{I}}(p, p'; q_1, q_2)$$
$$\times\, \ln[f(p)f(p')][g(q_1)g(q_2) - f(p)f(p')]$$
$$\times\, 2\, \mathrm{d}P\, \mathrm{d}P'\, \mathrm{d}Q_1\, \mathrm{d}Q_2. \tag{5.17}$$

Hence we see that $S_{\mathrm{L}}{}^{\mu}$ by itself, does not obey the H-theorem. This, again, is no surprise since the two components of the gas can exchange entropy; continuing the argument we see, in an obvious notation, that

$$S^{\mu}{}_{;\mu} = S_{\mathrm{L}}{}^{\mu}{}_{;\mu} + S_{\mathrm{I}}{}^{\mu}{}_{;\mu} > 0. \tag{5.18}$$

We would like, as in the case of the one component gas, to explore an iterative solution to the Boltzmann equations, expanding around the quasi-equilibrium solutions, the solutions which make $S^{\mu}{}_{;\mu}$ vanish. There are three conditions; i.e.,

$$\ln f_{\mathrm{eq}}(p) + \ln f_{\mathrm{eq}}(p') = \ln f_{\mathrm{eq}}(p_1) + \ln f_{\mathrm{eq}}(p_2) \tag{5.19}$$

$$\ln g_{\mathrm{eq}}(q) + \ln g_{\mathrm{eq}}(q') = \ln g_{\mathrm{eq}}(q_1) + \ln g_{\mathrm{eq}}(q_2) \tag{5.20}$$

and

$$\ln g_{\mathrm{eq}}(q_1) + \ln g_{\mathrm{eq}}(q_2) = \ln f_{\mathrm{eq}}(p) + f_{\mathrm{eq}}(p'), \tag{5.21}$$

where, in each case, the four momenta are constrained by an appropriate δ-function condition. The first two constraint equations can be satisfied by the choices

$$\ln f_{\text{eq}}(p) = \alpha_{\text{L}} - \beta_{\text{L}} E_{\text{L}}(p) \tag{5.22}$$

and

$$\ln g_{\text{eq}}(q) = \alpha_{\text{l}} - \beta_{\text{l}} E_{\text{l}}(q). \tag{5.23}$$

But to satisfy the constraint given by (5.21) we must have

$$\alpha_{\text{l}} = \alpha_{\text{L}} \tag{5.24}$$

and

$$\beta_{\text{L}} = \beta_{\text{l}}, \tag{5.25}$$

i.e., complete chemical and thermal equilibrium. Whether one chooses to expand about complete or partial equilibrium as expressed by (5.22) and (5.23) depends, of course, on the physical situation. These constraints along with the discussion following (3.67) raise an interesting question (see also Weinberg, 1971). Suppose we have a two-component gas one of whose components is massless and the other not. Is it possible for these components to be in chemical and thermal equilibrium? The answer is no, because that would mean for the massive component described by the putative equilibrium distribution f_{eq} that $L(f_{\text{eq}}) = 0$, and this we know is impossible. Hence the two components will, necessarily, "infect" each others distributions. This question arises when one asks how can the present-day ambient Big Band radiation be Planckian to such a high degree. This radiation last interacted with electrons at temperatures very much less than the electron mass. Why then did these mass effects not "infect" the photon distribution in a measurable way? We will come back to this in a later section, but the essential reason is that the photons outnumbered the electrons by a factor of something like a billion.

We shall next consider the expansions around the states of partial equilibrium; i.e., let

$$f = f_0(1 + \phi_{\text{L}}) \tag{5.26}$$

with

$$f_0 = \mathrm{e}^{\alpha_{\text{L}} - \beta_{\text{L}} E_{\text{L}}} \tag{5.27}$$

and

$$g = g_0(1 + \phi_{\text{l}}) \tag{5.28}$$

with

$$g_0 = e^{\alpha_1 - \beta_1 E_1}. \tag{5.29}$$

The interesting quantity to consider is

$$\begin{aligned} f(p)f(p') - g(q_1)g(q_2) \simeq\ & f_0(p)f_0(p') - g_0(q_1)g_0(q_2) \\ & + f_0(p)f_0(p')[\phi_L(p) + \phi_L(p')] \\ & - g_0(q_1)g_0(q_2)[\phi_l(q_1) + \phi_l(q_2)]. \end{aligned} \tag{5.30}$$

With this approximation we can write, for example,

$$\begin{aligned} \frac{\partial f}{\partial t} - \frac{\dot{R}}{R} p \frac{\partial f}{\partial p} = & \frac{1}{E(p)_L} \iiint (2\pi)^4 \delta^{(4)}(p + p' - p_1 - p_2) \\ & \times W_E^L(p, p'; p_1, p_2) f_0(p_1) f_0(p_2) \\ & \times [\phi_L(p_1) + \phi_L(p_2) - \phi_L(p) - \phi_L(p')]\, dP'\, dP_1\, dP_2 \\ & - \frac{1}{E(p)_L} \iiint (2\pi)^4 \delta^{(4)}(p + p' - q_1 - q_2) \\ & \times W_l(p, p'; q_1, q_2) \\ & \times \{[f_0(p)f_0(p') - g_0(q_1)g_0(q_2)] \\ & + f_0(p)f_0(p')[\phi_L(p) + \phi_L(p')] \\ & - g_0(q_1)g_0(q_2)[\phi_l(q_1) + \phi_l(q_2)]\}\, dP'\, dP_1\, dP_2 \end{aligned} \tag{5.31}$$

while,

$$\begin{aligned} \frac{\partial g}{\partial t} - \frac{\dot{R}}{R} q \frac{\partial g}{\partial q} = & \frac{1}{E(q)_l} \iiint (2\pi)^4 \delta^{(4)}(q + q' - q_1 - q_2) \\ & \times W_E^l(q, q'; q_1, q_2) g_0(q_1) g_0(q_2) \\ & \times [\phi_l(q_1) + \phi_l(q_2) - \phi_l(q') - \phi_l(q)]\, dQ_1\, dQ_2\, dQ' \\ & + \frac{1}{E(q)_l} \iiint (2\pi)^4 \delta^{(4)}(q + q' - p_1 - p_2) \\ & \times W_l(q, q'; p_1, p_2)\{[g_0(q)g_0(q') - f_0(p_1) f_0(p_2)] \\ & + g_0(q)g_0(q')[\phi_l(q) + \phi_l(q')] \\ & - f_0(p_1)f_0(p_2)[\phi_L(p_1) + \phi_L(p_2)]\}\, dQ'\, dP_1\, dP_2. \end{aligned} \tag{5.32}$$

Before discussing the physics of these equations we shall write out the expression for the entropy divergence; i.e.,

$$
\begin{aligned}
S^{\mu}{}_{;\mu} = &-\tfrac{1}{4}\iiiint (2\pi)^4\delta^{(4)}(p + p' - p_1 - p_2) \times W_{\mathrm{E}}^{\mathrm{L}}(p, p'; p_1, p_2)\\
&\times \ln\left(\frac{[1 + \phi_{\mathrm{L}}(p)][1 + \phi_{\mathrm{L}}(p')]}{[1 + \phi_{\mathrm{L}}(p_1)][1 + \phi_{\mathrm{L}}(p_2)]}\right) f_0(p_1) f_0(p_2)\\
&\times \{[1 + \phi_{\mathrm{L}}(p_1)][1 + \phi_{\mathrm{L}}(p_2)] - [1 + \phi_{\mathrm{L}}(p)][1 + \phi_{\mathrm{L}}(p')]\}\\
&\times 2\,\mathrm{d}P\,\mathrm{d}P'\,\mathrm{d}P_1\,\mathrm{d}P_2\\
&-\tfrac{1}{4}\iiiint (2\pi)^4\delta^{(4)}(q + q' - q_1 - q_2) \times W_{\mathrm{E}}^{\mathrm{l}}(q, q'; q_1, q_2)\\
&\times \ln\left(\frac{[1 + \phi_{\mathrm{l}}(q)][1 + \phi_{\mathrm{l}}(q')]}{[1 + \phi_{\mathrm{l}}(q_1)][1 + \phi_{\mathrm{l}}(q_2)]}\right) g_0(q_1) g_0(q_2)\\
&\times \{[1 + \phi_{\mathrm{l}}(q_1)][1 + \phi_{\mathrm{l}}(q_2)] - [1 + \phi_{\mathrm{l}}(q)][1 + \phi_{\mathrm{l}}(q')]\}\\
&\times 2\,\mathrm{d}Q\,\mathrm{d}Q'\,\mathrm{d}Q_1\,\mathrm{d}Q_2\\
&-\tfrac{1}{4}\iiiint (2\pi)^4\delta^{(4)}(p + p' - q_1 - q_2) W_{\mathrm{l}}(p, p'; q_1, q_2)\\
&\times \ln\left(\frac{f_0(p) f_0(p')[1 + \phi_{\mathrm{L}}(p)][1 + \phi_{\mathrm{L}}(p')]}{g_0(q_1) g_0(q_2)[1 + \phi_{\mathrm{l}}(q_1)][1 + \phi_{\mathrm{l}}(q_2)]}\right)\\
&\times \{g_0(q_1) g_0(q_2)[1 + \phi_{\mathrm{l}}(q_1)][1 + \phi_{\mathrm{l}}(q_2)]\\
&- f_0(p) f_0(p')[1 + \phi_{\mathrm{L}}(p)][1 + \phi_{\mathrm{L}}(p')]\} 2\,\mathrm{d}P\,\mathrm{d}P'\,\mathrm{d}Q_1\,\mathrm{d}Q_2\\
\simeq &\tfrac{1}{4}\iiiint (2\pi)^4\delta^{(4)}(p + p' - p_1 - p_2) W_{\mathrm{E}}^{\mathrm{L}}(p, p'; p_1, p_2)\\
&\times f_0(p_1) f_0(p_2)[\phi_{\mathrm{L}}(p_1) + \phi_{\mathrm{L}}(p_2) - \phi_{\mathrm{L}}(p) - \phi_{\mathrm{L}}(p')]^2\\
&\times 2\,\mathrm{d}P\,\mathrm{d}P'\,\mathrm{d}P_1\,\mathrm{d}P_2\\
&-\tfrac{1}{4}\iiiint (2\pi)^4\delta(q + q' - q_1 - q_2) W_{\mathrm{E}}^{\mathrm{l}}(q, q'; q_1, q_2)\\
&\times g_0(q_1) g_0(q_2)[\phi_{\mathrm{l}}(q_1) + \phi_{\mathrm{l}}(q_2) - \phi_{\mathrm{l}}(q) - \phi_{\mathrm{l}}(q')]^2\\
&\times 2\,\mathrm{d}Q\,\mathrm{d}Q'\,\mathrm{d}Q_1\,\mathrm{d}Q_2\\
&-\tfrac{1}{4}\iiiint (2\pi)^4\delta^{(4)}(p + p' - q_1 - q_2) W_{\mathrm{l}}(p, p'; q_1, q_2)\\
&\times \left[\ln\left(\frac{f_0(p) f_0(p')}{g_0(q_1) g_0(q_2)}\right) + [\phi_{\mathrm{L}}(p) + \phi_{\mathrm{L}}(p') - \phi_{\mathrm{l}}(q_1) - \phi_{\mathrm{l}}(q_2)]\right]\\
&+ \{g_0(q_1) g_0(q_2) - f_0(p) f_0(p') + g_0(q_1) g_0(q_2)[\phi_{\mathrm{l}}(q_1) + \phi_{\mathrm{l}}(q_2)]\\
&+ f_0(p) f_0(p')[\phi_{\mathrm{L}}(p) + \phi_{\mathrm{L}}(p')]\} 2\,\mathrm{d}P\,\mathrm{d}P'\,\mathrm{d}Q_1\,\mathrm{d}Q_2. \qquad (5.33)
\end{aligned}
$$

We see, from this rather complicated expression, that, in general, this expanding gas generates entropy to *first order* in ϕ. The special case in which this does not happen is when the two components are in chemical and thermal equilibrium so that equations (5.24) and (5.25) are satisfied. In this case we can repeat, to order ϕ^2, the analysis of Chapter 4, (4.21) *et seq.* That is, let

$$f_0 = Ae^{-\beta E_L}, \tag{5.34}$$

$$g_0 = Ae^{-\beta E_l}. \tag{5.35}$$

We have a gauge ambiguity in the ϕs which we can resolve by insisting that N_l, N_L and ρ_l, ρ_L, be given in terms of g_0 and f_0, just as we did for the one-component gas. Invoking the conservation of the total number $N_l + N_L$ and the approximate conservation of entropy we have, in this case, with U_l and U_L the total energies of the l particles and L particles,

$$\frac{\dot{A}}{A} = -3\frac{\dot{R}}{R} + \frac{\dot{\beta}}{N}(U_l + U_L) \tag{5.36}$$

and

$$\frac{\dot{A}}{A} = \frac{\partial}{\partial t}\frac{[\beta(U_l + U_L)]}{N} \tag{5.37}$$

or

$$\beta\frac{\dot{U}}{N} = -\frac{\dot{R}}{R}, \tag{5.38}$$

where†

$$U = U_l + U_L = R^3\left(\frac{3n_l}{\beta} + m_L n_L + \tfrac{3}{2}\frac{n_l}{\beta}\right). \tag{5.39}$$

To illustrate how complicated the temperature dependence on R can be even in this very simple case, let us suppose the l particle is massless and the temperature regime is such that $T \ll m_L$. Then (5.36) becomes

$$\beta\frac{\partial}{\partial t}\left(\frac{3N_l}{N\beta} + m_L\frac{N_L}{N} + \frac{3}{2}\frac{N_L}{N\beta}\right) = -3\frac{\dot{R}}{R}. \tag{5.40}$$

We see that, even in this very simple case, β may be a complicated function of R. To get a sense of how it behaves in various regimes let us take a simple

† This equation, and (4.32), are manifestations of the conservation of $T^{\mu\nu}$ conjoined with the equation of state of an ideal gas.

albeit artificial model that relates N_l and N_L; i.e., let

$$\dot{N}_L = -\Gamma N_L \tag{5.41}$$

and

$$\dot{N}_l = \Gamma N_L,$$

to be solved with the conditions that

$$N_L(0) = N_L + N_l \equiv N \tag{5.42}$$

and

$$N_l(0) = 0. \tag{5.43}$$

Thus

$$N_L = N e^{-\Gamma t}, \tag{5.44}$$

and

$$N_l = N(1 - e^{-\Gamma t}). \tag{5.45}$$

After some manipulation we find that for this model

$$\frac{\dot{\beta}}{\beta}\left(1 - \frac{N_L}{2N}\right) - \Gamma\frac{N_L}{N}(\tfrac{1}{2} - \tfrac{1}{3}\beta m_L) = \frac{\dot{R}}{R}. \tag{5.46}$$

Let us rewrite this equation, using (5.44), in the following form:

$$\frac{\dot{\beta}}{\beta} = \frac{(\dot{R}/R) - \Gamma e^{-\Gamma t}(\tfrac{1}{3}\beta m_L - \tfrac{1}{2})}{1 - \tfrac{1}{2}e^{-\Gamma t}}. \tag{5.47}$$

In principle, we would like to solve this equation to find β as a function of R and, eventually, of time. This requires knowing

$$\frac{\dot{R}}{R} = \frac{(\tfrac{8}{3}\pi\rho)^{\frac{1}{2}}}{M_{pl}}; \tag{5.48}$$

i.e., we must specify the cosmodynamics during the epoch of $L - \bar{L}$ annihilation. This we shall not try to do for this model problem, but rather, comment on some of the general features of (5.47). In the first place we see that for any Γ so long as $\Gamma t \gg 1$ we have

$$\frac{\dot{\beta}}{\beta} \simeq \frac{\dot{R}}{R}. \tag{5.49}$$

This is entirely to be expected. It says that after a time, long compared to the decay rate, the temperature dependence of the sensibly massless l particles, i.e., $TR \simeq$ constant, prevails. Furthermore if we set $\Gamma = 0$ we see we have

$$TR^2 \simeq \text{constant},$$

which is the temperature dependence of the massive L particles alone. The interesting physics emerges if we set $t = 0$ in (5.47). Thus

$$\left.\frac{\dot{\beta}}{\beta}\right|_{t=0} = \left.\frac{\dot{R}}{2R}\right|_{t=0} - \frac{\Gamma}{2}\left(\frac{\beta(0)m_L}{3} - \tfrac{1}{2}\right) \tag{5.50}$$

Therefore the sign of $\dot{\beta}/\beta|_{t=0}$ (remember that we are working in a regime with $\beta m_L > 1$) is fixed by the relative magnitudes of $\dot{R}/R$ and Γ. In the usual cosmological applications, where these considerations are applied, we have $\Gamma \gg \dot{R}/R$. Hence, initially, $\dot{\beta}/\beta$ is *negative*, which means that during the initial stages of the annihilation the temperature of the medium *increases*. When Γt becomes sensibly larger than one, the sign of the derivative changes and the temperature begins decreasing as $1/R$. The fact that annihilating particles heat the medium, in many cases, is familiar (see, for example, Weinberg, 1972). However, the usual treatment views this process as a series of discontinuous steps, like a jerky elevator, whereas it is clear from (5.47) that it is a perfectly continuous process. It is also interesting to point out that we could have $\Gamma > \dot{R}/R$ and $\dot{\beta}/\beta$ *positive*, if Γ is not too large. This means that there could be situations in which the interaction rate is too slow to allow the medium to be heated before the expansion cools it off. Each situation has to be examined before one can be confident that annihilating particles do heat the medium.†

Before turning to the first of the cosmologically interesting Lee–Weinberg problems, it is instructive to compute the total constant entropy, given our assumptions. To this end we employ (4.14) in the form

$$R^3 s = \beta U + N - N \ln A, \tag{5.51}$$

where A is the solution to (5.37). Hence, in this case,

$$R^3 s = N, \tag{5.52}$$

and the conservation of the total entropy is simply the conservation of the total particle number.

† If we had allowed somewhat more general initial conditions in the model so that $N_1(0) \neq 0$, i.e., some 1 particles were present in the medium before annihilation, (5.50) becomes,

$$\left.\frac{\dot{\beta}}{\beta}\right|_{t=0} = \frac{\dot{R}/R|_{t=0} - (\tfrac{1}{3}\beta m_L - \tfrac{1}{2})\Gamma N_L(0)/N}{[1 - \tfrac{1}{2}N_L(0)/N]},$$

which changes the analysis quantitatively but not qualitatively. The shortcoming of the model, as we shall see in the next chapter, is that it ignores recombination.

6
The generalized Lee–Weinberg problem: examples

The model problem we treated at the end of the last chapter was solved under the assumption that the L and l components were in both chemical and thermal equilibrium so that the Boltzmann functions were close to their equilibrium values f_0 and g_0. While this may be a good approximation at the beginning of the annihilation process, the fact that the energy barrier keeps the l particles from recombining into L particles soon pulls the L distribution away from its equilibrium value. It is this phenomenon that we wish to study quantitatively in this chapter. The treatment we will follow is the one given in Bernstein, Brown, and Feinberg (1985). A related and independent treatment can be found in Scherrer and Turner (1986). The first example we will study is the original Lee–Weinberg problem (1977) in which L is taken to be a very heavy neutral lepton – a heavy "neutrino" – and l is a light conventional neutrino. We shall begin by writing the equation for f, the Boltzmann function for each spin degree of freedom. Thus

$$\frac{\partial f}{\partial t} - \frac{\dot{R}}{R} p \frac{\partial f}{\partial p} = C_{\mathrm{E}}^{\mathrm{L}} + C_{\mathrm{I}}^{\mathrm{L}}, \tag{6.1}$$

and for g

$$\frac{\partial g}{\partial t} - \frac{\dot{R}}{R} q \frac{\partial g}{\partial q} = C_{\mathrm{E}}^{\mathrm{l}} + C_{\mathrm{I}}^{\mathrm{l}}, \tag{6.2}$$

where, invoking the symmetries and including the Fermi blocking factors such as $1 - f(p)$,

$$\begin{aligned} C_{\mathrm{E}}^{\mathrm{L}} = \frac{1}{E(p)_{\mathrm{L}}} \iiint & (2\pi)^4 \delta^{(4)}(p + q - p' - q') W_{\mathrm{E}}(p, q; p', q') \\ & \times \{[1 - f(p)][1 - g(q)] f(p') g(q') \\ & - [1 - f(p')][1 - g(q')] f(p) g(q)\} \, \mathrm{d}P' \, \mathrm{d}Q \, \mathrm{d}Q'. \end{aligned} \tag{6.3}$$

There might also be other elastic processes involving the L particles which we have not written explicitly. On the other hand,

$$C_{\mathrm{I}}^{\mathrm{L}} = \frac{1}{E(p)_{\mathrm{L}}} \iiint (2\pi)^4 \delta^{(4)}(p + \bar{p} - q - \bar{q}) W_{\mathrm{I}}(p, \bar{p}; q, \bar{q})$$
$$\times \{[1 - f(p)][1 - f(\bar{p})]g(q)g(\bar{q}) - [1 - g(q)][1 - g(\bar{q})]f(p)f(\bar{p})\}$$
$$\times \mathrm{d}\bar{P}\,\mathrm{d}Q\,\mathrm{d}\bar{Q}. \tag{6.4}$$

Concealed in (6.3) is an assumption, namely, although L and $\bar{\mathrm{L}}$, the L antiparticle, may be distinct, and likewise so may l and $\bar{\mathrm{l}}$, we assume that the scattering and annihilation processes are symmetric between particle and antiparticle. Hence both L and $\bar{\mathrm{L}}$ are described by the same function f and, likewise, l and $\bar{\mathrm{l}}$ are described by the same function g. It is clear from the structure involving the fs and gs in (6.3) that it changes sign if we exchange p with p', and q with q'. Hence, with our usual assumptions about the symmetries of W_{E}, we have

$$\int C_{\mathrm{E}}^{\mathrm{L}}(p)\,\mathrm{d}^3p = 0. \tag{6.5}$$

Hence, if we wish to study the number density of L particles or $\bar{\mathrm{L}}$ particles as a function of time, as opposed to the time development of the fs, these elastic terms drop out of the equation, and we have

$$\frac{1}{R^3}\frac{\partial}{\partial t}\left(R^3 \int f \frac{\mathrm{d}^3p}{(2\pi)^3}\right) = \int C_{\mathrm{I}}^{\mathrm{L}}(p)\frac{\mathrm{d}^3p}{(2\pi)^3}. \tag{6.6}$$

Since L is a spin-$\frac{1}{2}$ particle, in equilibrium each of its two spin components, all things being equal, contributes equally to the number density. Thus, it is natural to call the number density

$$n(t) = 2\int f \frac{\mathrm{d}^3p}{(2\pi)^3}. \tag{6.7}$$

The same considerations apply to the l number density but we will not be interested in determining it explicitly, so there will be no confusion if we do not label n with an "L". Hence the equation to be solved is

$$\frac{1}{R^3}\frac{\partial}{\partial t}[R^3 n(t)] = 4 \iiiint (2\pi)^4 \delta^{(4)}(p + \bar{p} - q - \bar{q})$$
$$\times W_{\mathrm{I}}(p, \bar{p}; q, q') \times \{[1 - f(p)][1 - f(\bar{p})]g(q)g(\bar{q})$$
$$- [1 - g(q)][1 - g(\bar{q})]f(p)f(\bar{p})\}\,\mathrm{d}P\,\mathrm{d}\bar{P}\,\mathrm{d}Q\,\mathrm{d}\bar{Q}. \tag{6.8}$$

This equation cannot even be solved in principle without knowing and solving the related equation for g. Here we are saved by the physics. In the situations we are interested in, the annihilation products are kept in tight equilibrium by scattering reactions involving each other and other particles and radiation. The use of the equilibrium distribution raises an important question of principle involving the chemical potentials. If we examine the argument that led to (5.24) and (5.25) it is clear that if we have a group of particles A, B, C... which enter a reaction which produces particles A′, B′, C′... and if these particles are in chemical equilibrium then we must have,

$$\alpha_{\mathrm{A}} + \alpha_{\mathrm{B}} + \alpha_{\mathrm{C}} + \cdots = \alpha_{\mathrm{A}'} + \alpha_{\mathrm{B}'} + \alpha_{\mathrm{C}'} + \cdots, \tag{6.9}$$

where the αs are the chemical potentials of the particles in question. This enables us to draw some conclusions about the chemical potentials of various particles in chemical equilibrium.

(1) The chemical potential of the photon, α_γ, is necessarily zero. This is traceable to the fact that the photon number is not conserved. Specifically a reaction like

$$\mathrm{e} + \mathrm{e} \rightarrow \mathrm{e} + \mathrm{e} + \gamma,$$

electron – electron bremsstrahlung, would violate (6.9) unless α_γ were zero.

(2) From the vanishing of α_γ it is clear that if we can have a reaction like

$$\mathrm{L} + \bar{\mathrm{L}} \rightarrow \gamma + \gamma,$$

we must have

$$\alpha_{\mathrm{L}} = -\alpha_{\bar{\mathrm{L}}}. \tag{6.10}$$

(3) Suppose there is an L described approximately by an equilibrium distribution of the form, taking into account the Fermi–Dirac statistics,

$$f_0^{\mathrm{L}} = 1/[\exp(\alpha_{\mathrm{L}} + \beta E_{\mathrm{L}}) + 1]. \tag{6.11}$$

Then, from (2),

$$f_0^{\bar{\mathrm{L}}} = 1/[\exp(-\alpha_{\mathrm{L}} + \beta E_{\mathrm{L}}) + 1]. \tag{6.12}$$

Hence, unless α_{L} is *zero*,

$$N_{\mathrm{L}} \neq N_{\bar{\mathrm{L}}}.$$

For some L particles, such as the electron, we know from charge neutrality that, in the early universe, prior to electron – positron annihilation, $N_{\mathrm{e}^-} \simeq N_{\mathrm{e}^+}$. Hence we shall take $\alpha_{\mathrm{e}^-} = 0$. The neutrino chemical potential is a much more elusive quantity (see, for example, Weinberg, 1972) and, at least in principle, could be quite large in some regimes. In the examples we shall consider we shall, in the absence of a reason to the contrary, set this chemical potential equal to zero. Therefore in solving the equation for f, or n, we will take

$$g = g_0 = 1/(e^{\beta E_l} + 1). \tag{6.13}$$

After we solve for f we could, in the spirit of an iteration procedure, use this f to find a corrected g, and so on. However, as is well-known, solving the simplest Boltzmann integro-differential equation is not trivial. Hence it is not obvious how to carry out even the first step in this program. To parametrize our difficulties we introduce a specific trial form for f; i.e.,

$$f = \frac{1}{\exp[\alpha(t) + \phi(p,t) + \beta E_L] + 1}. \tag{6.14}$$

Several comments are in order about this f. In the first place, comparison with (6.13) shows that we are assuming thermal equilibrium between the l particles and L particles. In the physical situations we want to consider the l particles and L particles are in an ambient bath of particles such as quarks, electrons, and photons and it is the contact with this reservoir that determines their common temperature. In the second place, we must explain the different roles of the ϕ and α. Even if there were no annihilation then, as we have seen, it would still not be strictly correct to take for f its equilibrium form. It is the ϕ that compensates. We are supposing that elastic scattering is intense and hence that the ϕ corrections are small. Despite appearances, α is not a chemical potential. It is a parameter which we take to be the *same* for L and $\bar{\mathrm{L}}$. The utility of this will become clear in the sequel. We may call α a "pseudo chemical potential" and this method of solving the coupled Boltzmann equations the method of "pseudo chemical potentials."

We begin by defining f_0 as

$$f_0 = \frac{1}{\exp[\alpha(t) + \beta E_L] + 1}. \tag{6.15}$$

Thus we write

$$f = f_0[1 + [1 - f_0]\phi] + O(\phi)^2. \tag{6.16}$$

The classical case we considered earlier – Chapter 5 – corresponded to having $f_0 \ll 1$. We now argue that ϕ is to be determined by C_E. In fact if f is simply inserted bodily into (6.1) we see that there will be terms linear in ϕ on both sides of the equation. What distinguishes the ϕ terms in C_{I} from those in C_{E} is that they will, essentially, be multiplied by n^2 as opposed to n. But, because of the annihilation, n is a rapidly decreasing function of time. Hence we can neglect $n^2\phi$ terms in C_{I} compared to $n\phi$ terms in C_{E}. Thus ϕ is, sensibly, determined by the elastic collisions alone, which we assume are very strong, making $\phi \ll 1$. Hence the process is now effectively parametrized in terms of the single parameter $\alpha(t)$. With these physically plausible approximations (6.8) can now be written

$$\frac{1}{R^3}\frac{\partial}{\partial t}(R^3 n) = 4\iiiint (2\pi)^4\delta^{(4)}(p + \bar{p} - q - \bar{q}) \times W_{\mathrm{I}}(p, \bar{p}; q, \bar{q})[\exp(2\alpha) - 1] \times f_0(p) f_0(\bar{p})[1 - g_0(q)][1 - g_0(\bar{q})]\,\mathrm{d}P\,\mathrm{d}\bar{P}\,\mathrm{d}Q\,\mathrm{d}\bar{Q}. \quad (6.17)$$

It is clear from the structure of (6.17) that it is the departure of α from zero that measures the loss of L particles through annihilation. Although we are concentrating on the annihilation of L particles, it follows at once with the same approximations that the creation of l particles is governed by the equation, invoking the symmetries

$$\frac{1}{R^3}\frac{\partial}{\partial t}(R^3 n_1) = -4\iiiint (2\pi)^4\delta^{(4)}(p + \bar{p} - q - \bar{q}) \times W_{\mathrm{I}}(p, \bar{p}; q, \bar{q}) \times (\exp(2\alpha) - 1) \times f_0(p) f_0(\bar{p})[1 - g_0(q)][1 - g_0(\bar{q})]\,\mathrm{d}P\,\mathrm{d}\bar{P}\,\mathrm{d}Q\,\mathrm{d}\bar{Q}, \quad (6.18)$$

with the *same* α, so that

$$(n + n_1)^{\mu}{}_{;\mu} = 0. \quad (6.19)$$

Even these equations are very difficult to solve. Here, once again, the physics enables us to simplify. The annihilation becomes significant only when $T < m_{\mathrm{L}} \leqslant E_{\mathrm{L}}$, and when $\alpha(t) > 1$. Thus, during this regime, it is a good approximation to use

$$f_0(p) = 1/[\exp(\alpha + \beta E_{\mathrm{L}}) + 1] \simeq \exp(-\alpha - \beta E_{\mathrm{L}}). \quad (6.20)$$

Thus

$$n(t) \simeq \mathrm{e}^{-\alpha(t)} n_{\mathrm{cl}}(t), \quad (6.21)$$

where†

$$n_{\mathrm{cl}}(t) = 2\int \mathrm{e}^{-\beta E_{\mathrm{L}}}\frac{\mathrm{d}^3p}{(2\pi)^3}.$$

$$\simeq \frac{1}{2\pi^2}\,\mathrm{e}^{-\beta m_{\mathrm{L}}}\,\frac{m_{\mathrm{L}}(2\pi m_{\mathrm{L}})^{\frac{1}{2}}}{\beta^{\frac{3}{2}}}. \tag{6.22}$$

In (6.18) the f_0 functions are integrated over the momenta. These integrands are restricted by the magnitude of, say, $\mathrm{e}^{-E(p)_{\mathrm{L}}\beta}$, where in this regime,

$$\mathrm{e}^{-E(p)_{\mathrm{L}}\beta} \simeq \mathrm{e}^{-m_{\mathrm{L}}\beta}\mathrm{e}^{-p^2\beta/2m_{\mathrm{L}}}. \tag{6.23}$$

This is substantial only if

$$p \lesssim (2m_{\mathrm{L}}/\beta)^{\frac{1}{2}}. \tag{6.24}$$

Furthermore, in this regime,

$$E(p)_{\mathrm{L}} + E(\bar{p})_{\mathrm{L}} \simeq 2m_{\mathrm{L}}.$$

Thus the energy-conserving δ-function is given approximately by

$$\delta[E(p)_{\mathrm{L}} + E(\bar{p})_{\mathrm{L}} - E(q)_1 - E(\bar{q})_1] \simeq \delta(2m_{\mathrm{L}} - q - \bar{q}). \tag{6.25}$$

This means that

$$q \simeq \bar{q} \simeq m_{\mathrm{L}} > p, \tag{6.26}$$

and therefore,

$$\delta^{(3)}(\mathbf{p} + \bar{\mathbf{p}} - \mathbf{q} - \bar{\mathbf{q}}) \simeq \delta^{(3)}(\mathbf{q} + \bar{\mathbf{q}}). \tag{6.27}$$

We shall use these relations to simplify the collision term. Moreover, since $q \simeq m_{\mathrm{L}}$ and $T < m_{\mathrm{L}}$, we can write

$$1 - g_0(q) = \frac{\mathrm{e}^{\beta E(q)_1}}{1 + \mathrm{e}^{\beta E(q)_1}} \simeq 1. \tag{6.28}$$

Thus the inelastic collision term can be simplified as follows:

$$C_{\mathrm{I}}^{\mathrm{L}}(p) = \frac{1}{E(p)_{\mathrm{L}}}\iiint (2\pi)^4\delta^{(4)}(p + \bar{p} - q - \bar{q})$$
$$\times\, W_{\mathrm{I}}(p,\bar{p};q,\bar{q})\{[1 - f(p)][1 - f(\bar{p})]g(q)g(\bar{q})$$
$$-\,[1 - g(q)][1 - g(\bar{q})]f(p)f(\bar{p})\}\,\mathrm{d}\bar{P}\,\mathrm{d}Q\,\mathrm{d}\bar{Q}$$

† This assumes that both L spin degrees of freedom are in equilibrium.

$$\simeq \frac{1}{E(p)_{\mathrm{L}}} \iiint (2\pi)^4 \delta(2m_{\mathrm{L}} - q - \bar{q}) \delta^{(3)}(\mathbf{q} + \bar{\mathbf{q}})$$

$$\times W_{\mathrm{I}}(p, \bar{p}; q, \bar{q})(1 - \mathrm{e}^{-2\alpha}) \mathrm{e}^{-\beta[E_{\mathrm{L}}(p) + E_{\mathrm{L}}(\bar{p})]} \, \mathrm{d}\bar{P} \, \mathrm{d}Q \, \mathrm{d}\bar{Q}. \quad (6.29)$$

For completeness, we also write the expression for $C_{\mathrm{E}}^{\mathrm{L}}(p)$ in which we cannot make use of the same kinematic approximations. Using the notation of (6.20) we can write

$$C_{\mathrm{E}}^{\mathrm{L}}(p) = \frac{1}{E(p)_{\mathrm{L}}} \iiint (2\pi)^4 \delta^{(4)}(p + q - p' - q')$$

$$\times W_{\mathrm{E}}(p, q; p', q') \mathrm{e}^{\alpha} (\mathrm{e}^{\beta[E(p)_{\mathrm{L}} + E(q)_{\mathrm{l}}]} - \mathrm{e}^{\beta[E(p')_{\mathrm{L}} + E(q')_{\mathrm{l}}]})$$

$$\times f_0(p) f_0(p') g_0(q) g_0(q') \, \mathrm{d}P' \, \mathrm{d}Q \, \mathrm{d}Q' = 0. \quad (6.30)$$

This last is a consequence of the conservation of energy. Likewise

$$C_{\mathrm{E}}^{\mathrm{l}}(q) = 0. \quad (6.31)$$

Hence with our *anzatz* f_0 is determined by the inelastic collisions. We will use this fact along with the expression for $C_{\mathrm{I}}^{\mathrm{l}}(q)$, which we write down next, to determine the entropy generated during the annihilation. Thus

$$C_{\mathrm{I}}^{\mathrm{l}}(q) = -\frac{1}{E(q)_{\mathrm{l}}} \iiint (2\pi)^4 \delta^{(4)}(p + \bar{p} - q - \bar{q})$$

$$\times W_{\mathrm{I}}(p, \bar{p}; q, \bar{q}) \{[1 - f(p)][1 - f(\bar{p})] g(q) g(\bar{q})$$

$$- [1 - g(q)][1 - g(\bar{q})] f(p) f(\bar{p})\} \, \mathrm{d}P \, \mathrm{d}\bar{P} \, \mathrm{d}\bar{Q}$$

$$\simeq -\frac{1}{E(q)_{\mathrm{l}}} \iiint (2\pi)^4 \delta[2m_{\mathrm{L}} - E(q)_{\mathrm{l}} - E(\bar{q})_{\mathrm{l}}]$$

$$\times \delta^{(3)}(\mathbf{q} + \bar{\mathbf{q}}) W_{\mathrm{I}}(p, \bar{p}; q, \bar{q})(1 - \mathrm{e}^{-2\alpha}) \mathrm{e}^{-\beta[E_{\mathrm{l}}(q) + E_{\mathrm{l}}(q)]} \, \mathrm{d}P \, \mathrm{d}\bar{P} \, \mathrm{d}\bar{Q}. \quad (6.32)$$

Now, for a Fermi–Dirac particle, with its entropy current defined by taking the lower signs in (3.55), with $C(E)$ the generic collision term, assuming two spin degrees of freedom,

$$S^{\mu}{}_{;\mu} = -2 \int C(E) \ln\left[\frac{f}{1 - f}\right] \frac{\mathrm{d}^3 p}{(2\pi)^3}. \quad (6.33)$$

Using the exact form of f_0 we have

$$\ln\left[\frac{f_0}{1 - f_0}\right] = -(\alpha + \beta E_{\mathrm{L}}), \quad (6.34)$$

which is the same answer we would have obtained if we had used the classical form of the f_0 in (6.20) and the approximate form of the corresponding logarithm. Thus we find for $S_L{}^\mu{}_{;\mu}$, using (6.32) and (6.34),

$$S_L{}^\mu{}_{;\mu} = \iiiint [\alpha + \beta E(p)] \times (2\pi)^4 \delta[2m_L - E(q)_1 - E(\bar{q})_1]\delta^{(3)}(\mathbf{q} + \bar{\mathbf{q}}) \times W_1(p, \bar{p}; q, \bar{q})(1 - e^{-2\alpha})e^{-\beta[E_L(p)+E_L(\bar{p})]} 4\,dP\,d\bar{P}\,dQ\,d\bar{Q}. \quad (6.35)$$

In the same spirit we can find $S_1{}^\mu{}_{;\mu}$ using the g_0 of (6.13). Thus

$$S_1{}^\mu{}_{;\mu} = -\iiiint \beta E(q)(2\pi)^4 \delta[2m_L - E(q)_1 - E(\bar{q})_1]\delta^{(3)}(\mathbf{q} + \bar{\mathbf{q}}) \times W_1(p, \bar{p}; q, \bar{q})(1 - e^{-2\alpha})e^{-\beta[E_L(p)+E_L(\bar{p})]} 4\,dP\,d\bar{P}\,dQ\,d\bar{Q}. \quad (6.36)$$

If we add (6.35) and (6.36) and use the symmetries and the conservation of energy we find for the total entropy S^μ,

$$S^\mu{}_{;\mu} = \iiiint \alpha(2\pi)^4 \delta[2m_L - E(q)_1 - E(\bar{q})_1] \times \delta^{(3)}(\mathbf{q} + \bar{\mathbf{q}}) W_1(p, \bar{p}; q, \bar{q}) \times (1 - e^{-2\alpha})e^{-\beta[E_L(p)+E_L(\bar{p})]} 4\,dP\,d\bar{P}\,dQ\,d\bar{Q}. \quad (6.37)$$

Before commenting on this expression let us rewrite it using the following definitions and identities. In the first place from (6.21) and (6.22) we have the obvious identity

$$1 - e^{-2\alpha} = (n_{cl}^2 - n^2)/n_{cl}^2. \quad (6.38)$$

Then, in the spirit of (4.41), we define $\langle \sigma v \rangle_0$ as

$$\langle \sigma v \rangle_0 = \iint (2\pi)^4 \delta(2m_L - q - \bar{q})\delta^{(3)}(\mathbf{q} + \bar{\mathbf{q}}) \times \frac{W_1(p, \bar{p}; q, \bar{q})}{4m_L^2}\,dQ\,d\bar{Q}. \quad (6.39)$$

The reason for the subscript on $\langle \sigma v \rangle_0$ will emerge in what follows. Thus we can write,

$$S^\mu{}_{;\mu} = \alpha(n_{cl}^2 - n^2)\langle \sigma v \rangle_0 = n\alpha\left[\frac{n_{cl}^2}{n^2} - 1\right] n\langle \sigma v \rangle_0. \quad (6.40)$$

We can now comment on this expression. From the second law we must have $S^{\mu}{}_{;\mu} > 0$. Referring to (6.37) this means that

$$\alpha(1 - e^{-2\alpha}) > 0.$$

This is satisfied for either α positive or negative. To determine the correct choice we invoke the rate equations (6.17) and (6.18). The L particles are decreasing during the annihilation regime while the l particles are increasing. A glance at these equations shows that α must be *negative.* As (6.40) makes clear, $S^{\mu}{}_{;\mu}$ is proportional to the annihilation rate $r\langle\sigma v\rangle_0$, as is to be expected. We can put this in a somewhat different way by rewriting the rate equation (6.17) using the same set of approximations that lead to (6.40). Thus we write,

$$\begin{aligned}\frac{1}{R^3}\left(\frac{\partial}{\partial t}(R^3 n)\right) &= 4\iiiint(2\pi)^4\delta^{(4)}(p + \bar{p} - q - \bar{q}) \\ &\quad\times W_l(p, \bar{p}; q, \bar{q})[\exp(2\alpha) - 1] \\ &\simeq (1 - e^{-2\alpha})n_{cl}^2\langle\sigma v\rangle_0 \\ &= (n_{cl}^2 - n^2)\langle\sigma v\rangle_0,\end{aligned} \tag{6.41}$$

where, in the last step, we have used (6.38). We may call this the "special" Lee–Weinberg equation (Lee and Weinberg 1977).[†] The "general" Lee–Weinberg equation is obtained by beginning with the first step in (6.41) and defining $\langle\sigma v\rangle$ as

$$\begin{aligned}\langle\sigma v\rangle = {} & \frac{1}{n_{cl}}\int\frac{d^3p}{(2\pi)^3}2e^{-\beta E_L(p)} \times \frac{1}{n_{cl}}\int\frac{d^3\bar{p}}{(2\pi)^3}2e^{-\beta E_L(\bar{p})} \\ & \times\iint(2\pi)^4\delta^{(4)}(p + \bar{p} - q - \bar{q})W_l(p, \bar{p}; q, \bar{q}) \\ & [1 - g_0(q)][1 - g_0(\bar{q})]\frac{dQ\,d\bar{Q}}{4E(p)E(\bar{p})}.\end{aligned} \tag{6.42}$$

Then the general equation is

$$\frac{1}{R^3}\left[\frac{\partial}{\partial t}(R^3 n)\right] = (n_{cl}^2 - n^2)\langle\sigma v\rangle. \tag{6.43}$$

[†] The use of a rate equation of the form (6.43) was well-known to astrophysicists prior to its use by Lee and Weinberg. Lee and Weinberg applied it to the heavy lepton problem. Since that is the problem we are dealing with, we have used this nomenclature. See Zel'dovich (1965) and Vysotskii, Dolgov, and Zel'dovich (1977).

Thus (6.40) can be written

$$S^{\mu}{}_{;\mu} = \alpha \frac{1}{R^3} \frac{\partial}{\partial t}(R^3 n). \tag{6.44}$$

This expression can also be obtained by using the conservation of the energy–momentum tensor and the total number current. Indeed, let us use these conditions to study the evolution of the temperature during the annihilation regime. Using the distributions of (6.20) and (6.28) we have for the partial pressures of the l and L components,†

$$P_l = n_l/\beta \tag{6.45}$$

and

$$P_L = n_L/\beta. \tag{6.46}$$

Now, using the conservation of the energy-momentum tensor, (3.42), we have in terms of the total energy U,

$$\beta\dot{U} = -3\frac{\dot{R}}{R}N, \tag{6.47}$$

where N is the total particle number. It is clear from this equation that

$$\dot{U} < 0. \tag{6.48}$$

To proceed, we write U as

$$U = 3(N_l/\beta) + m_L N_L + \tfrac{3}{2}N_L/\beta. \tag{6.49}$$

Thus, using the conservation of N,

$$\beta\dot{U} = (\beta m_L - \tfrac{3}{2})\frac{\partial N_L}{\partial t} - \frac{\dot{\beta}}{\beta}(3N_l + \tfrac{3}{2}N_L) \tag{6.50}$$

$$= -3\frac{\dot{R}}{R}N.$$

This equation exhibits the correct limiting cases. If we let $t \to \infty$, for example, $\partial N_L/\partial t \to 0$, since the L annihilations become less probable, and $N_L \to 0$. Thus, in this limit

$$\frac{\dot{\beta}}{\beta} = \frac{\dot{R}}{R} \tag{6.51}$$

† In what follows we always include under the rubric "L particle" the anti-L particles.

and the gas expands adiabatically with $T \sim 1/R$. On the other hand if the annihilation is shut off so that $\partial N_L/\partial t = 0$ and both N_l and N_L are constant we have, using (6.50)

$$\frac{1}{T} \sim R^{N/(N_l + N_L/2)} = R^{2/(1 + N_l/N)} \tag{6.52}$$

which says that in this limit the gas expands with T varying between $1/R$ and $1/R^2$ depending on the admixture of l and L particles. In any event, it is instructive to study the sign of $\dot{\beta}/\beta$ at $t = 0$. To this end we need the value of $\partial N_L/\partial t$ at $t = 0$. In the model example in the last chapter we set $(\partial N_L/\partial t)|_{t=0} = -\Gamma N_L(0)$, where Γ is the L–$\bar{\text{L}}$ annihilation rate. We may, in light of the more complete treatment in terms of a rate equation, ask what the region of validity of the model example is likely to be. Thus we can write (6.43) in the form,

$$\dot{N}_L = N_{cl} n_{cl} \langle \sigma v \rangle - N_L n \langle \sigma v \rangle. \tag{6.53}$$

In the limit of large t the first term will fall off more rapidly than the second – a consequence of recombination – and thus in this limit

$$\dot{N}_L \simeq -N_L \Gamma, \tag{6.54}$$

where

$$\Gamma = n\langle \sigma v \rangle. \tag{6.55}$$

Thus we expect the model to reflect large t behavior and not, except under special circumstances, the behavior at $t = 0$. However, we expect (6.53) to give an accurate representation of N_L for all t, including $t = 0$. In the L–l system we are considering, the particles are assumed to be in equilibrium at $t = 0$. This is made possible because of recombination, the effect that is left out in (6.54). Hence the correct initial condition is

$$N_L(0) = N_{cl}(0), \tag{6.56}$$

or

$$\dot{N}_L \Big|_{t=0} = 0. \tag{6.57}$$

Thus, from (6.50)

$$\frac{\dot{\beta}}{\beta}\Big|_{t=0} = \frac{\dot{R}}{R}\Big|_{t=0} \frac{N}{N_l(0) + \frac{1}{2}N_L(0)} > 0. \tag{6.58}$$

Thus, given our assumptions, the temperature *decreases* initially. To follow the behavior of the temperature as a function of time in detail we need both a model for $\dot{R}/R$ and the solution to (6.53). We turn to these matters after noting that if we had added to the particles present at $t = 0$ light quarks and leptons which obey a number conservation law this would not change the conclusion of (6.58). However, radiation, which does not obey such a conservation law, must be treated separately. We return to this when we discuss e^+–e^- annihilation.

We cannot even begin to solve the rate equation without having an expression for $\dot{R}/R$. Since

$$\left[\frac{\dot{R}}{R}\right]^2 = \frac{8\pi}{3}\frac{\rho}{M_{\mathrm{pl}}^2}, \tag{6.59}$$

this means, to determine $\dot{R}/R$, we must have an expression for ρ. Needless to say, the L and l particles contribute to ρ. If they were the only contributors, or even the dominant contributors, this would make the rate equation hopelessly complicated to solve since ρ would itself depend primarily on N_{L} and N_{l}. It is fortunate that during the regimes of interest ρ is dominated by other sensibly massless particles so that $\dot{R}/R$ becomes a known function of temperature. We have no reason to suppose that it is correct to ignore the quantum statistics for these particles, which we can take to be approximately in equilibrium. Thus for each such Fermi–Dirac degree of freedom in equilibrium

$$\begin{aligned}\rho_i^{\mathrm{FD}} &= \frac{1}{2\pi^2}\int_0^\infty p^3 \frac{1}{e^{p\beta}+1}\,\mathrm{d}p \\ &= \frac{1}{2\pi^2\beta^4}\int_0^\infty x^3 \frac{1}{e^x+1}\,\mathrm{d}x = \frac{7\pi^2}{240}\frac{1}{\beta^4},\end{aligned} \tag{6.60}$$

while for Bose–Einstein particles the corresponding expression is

$$\rho_i^{\mathrm{BE}} = \frac{1}{2\pi^2\beta^4}\int_0^\infty x^3\frac{1}{e^x-1}\,\mathrm{d}x = \frac{\pi^2}{30}\frac{1}{\beta^4}, \tag{6.61}$$

so that

$$\frac{\rho_i^{\mathrm{FD}}}{\rho_i^{\mathrm{BE}}} = \tfrac{7}{8}. \tag{6.62}$$

Thus, in this regime,

$$\rho = \rho_0 + \rho_1, \tag{6.63}$$

where ρ_0 represents the contribution of the massless particles in equilibrium and ρ_1, the rest. It is difficult to calculate ρ_1 since, as a glance at (6.49) shows, ρ_1 depends on the number of Ls or ls, which is what we are trying to compute. However, if we are working in a regime in which $\rho_0 \gg \rho_1$ we can, as a first approximation, ignore ρ_1 in computing $\dot{R}/R$. In general, ρ_0 is given by

$$\rho_0 = \frac{\pi^2}{30}\frac{1}{\beta^4}\left(\sum_i N_i^{\mathrm{BE}} + \tfrac{7}{8}\sum_i N_i^{\mathrm{FD}}\right), \tag{6.64}$$

where the sum is over the massless degrees of freedom in equilibrium. We shall specify these more precisely below. But let us assume that they are sufficiently numerous so that they outweigh ρ_1. Thus, noting that for any massless particle, see (3.37) and (3.39),

$$P = \tfrac{1}{3}\rho \tag{6.65}$$

we have, approximately, from the conservation of the energy–momentum tensor,

$$\frac{\partial}{\partial t}U = -\frac{\dot{R}}{R}U, \tag{6.66}$$

which we can solve to give, where C is a constant,

$$U = C/R. \tag{6.67}$$

Thus, using (6.64), we see that to a first approximation,

$$T = C'/R, \tag{6.68}$$

where C' is another constant. The strategy is now clear. We will solve the rate equation using this first approximation for $\dot{R}/R$ and then, having found N_{L}, we can use it, if we want, to examine the departures of the temperature from (6.68). Hence, putting together (6.64) and (6.68) we find, in this approximation,

$$(\dot{R}/R)^2 = (\dot{T}/T)^2 = AT^4, \tag{6.69}$$

where

$$A = \frac{8\pi}{3}\frac{1}{M_{\mathrm{pl}}^2}\frac{\pi^2}{30}N^{\mathrm{DF}} \tag{6.70}$$

and

$$N^{\mathrm{DF}} = \sum_i N_i^{\mathrm{BE}} + \tfrac{7}{8}\sum_i N_i^{\mathrm{FD}}. \tag{6.71}$$

Using these definitions, and the preceding equations, we have the important identity, valid in this regime,

$$\frac{\partial}{\partial t} = -T^3 A^{\frac{1}{2}}\frac{\partial}{\partial T}. \tag{6.72}$$

Hence the rate equation can be written in the form

$$\frac{d}{dT}\left(\frac{n(T)}{T^3}\right) = \frac{\langle\sigma v\rangle}{A^{\frac{1}{2}}}\left[\left(\frac{n(T)}{T^3}\right)^2 - \left(\frac{n(T)_{cl}}{T^3}\right)^2\right]. \tag{6.73}$$

To study the short time, or high temperature, behavior of α, and hence N_L, it is useful to write

$$n = e^{-\alpha}T^3G_0 \tag{6.74}$$

where

$$G_0 = \frac{1}{\pi^2}\int_0^\infty dx\, x^2 e^{-(x^2+m_L^2\beta^2)^{1/2}}. \tag{6.75}$$

Thus, in this language, (6.73) becomes

$$\frac{d}{dT}\left(e^{-\alpha}G_0\right) = \frac{\langle\sigma v\rangle}{A^{\frac{1}{2}}} G_0^2(e^{-2\alpha} - 1). \tag{6.76}$$

The boundary condition is that at $T = T_0$, $\alpha(0) = 0$, representing the initial equilibrium. It is useful, in discussing magnitudes, to introduce the dimensionless variable,

$$Y = T/m_L. \tag{6.77}$$

Thus (6.76) can be written

$$\frac{d}{dy}(e^{-\alpha}G_0) = \lambda G_0^2[e^{-2\alpha} - 1] \tag{6.78}$$

where λ is the dimensionless quantity

$$\lambda = \langle\sigma v\rangle m_L/A^{\frac{1}{2}}. \tag{6.79}$$

In all our examples we will discover that $\lambda \gg 1$. It is the magnitude of λ that determines how long the equilibrium is maintained once the L and $\bar{\text{L}}$ begin to annihilate. For that reason we anticipate that α goes to zero at least as rapidly as $1/\lambda$. For short times and large λ we can contemplate an expansion of the form

$$\frac{d}{dy}[(1-\alpha)G_0] \simeq -\lambda G_0^2 2\alpha. \tag{6.80}$$

Since $\alpha \sim 1/\lambda$, the α term on the left-hand side of (6.80) can be neglected and, in so far as the expansion is valid

$$\alpha \simeq -G_0'/G_0^2 2\lambda. \tag{6.81}$$

Since, written in terms of y,

$$G_0 = \frac{1}{\pi^2}\int_0^\infty dx\, x^2 e^{-(x^2+1/y^2)^{1/2}}, \qquad (6.82)$$

we have

$$G_0' = \frac{1}{\pi^2}\int_0^\infty \frac{dx\, x^2}{(x^2+1/y^2)^{\frac{1}{2}}}\,\frac{e^{-(x^2+1/y^2)^{1/2}}}{y^3} > 0, \qquad (6.83)$$

and α is negative as advertized. We expect that α will become large only when $T/m \ll 1$. In this regime we can estimate G_0; i.e., for $y \ll 1$,

$$G_0 \simeq \frac{1}{2^{\frac{1}{2}}\pi^{\frac{3}{2}}}\frac{1}{y^{\frac{3}{2}}}e^{-1/y}. \qquad (6.84)$$

It is also useful to evaluate G_0 in the limit $y \gg 1$; i.e.,

$$G_0 \simeq \frac{2}{\pi^2}\left(1 - \frac{1}{4y^2}\right). \qquad (6.85)$$

The derivatives in the corresponding limits are: for $y \ll 1$

$$G_0' \simeq \frac{1}{2^{\frac{1}{2}}\pi^{\frac{3}{2}}}e^{-1/y}\left(\frac{1}{y^{\frac{7}{2}}} - \frac{3}{2}\frac{1}{y^{\frac{5}{2}}}\right)$$

$$\simeq \frac{1}{2^{\frac{1}{2}}\pi^{\frac{3}{2}}}e^{-1/y}\frac{1}{y^{\frac{7}{2}}}, \qquad (6.86)$$

while for $y \gg 1$,

$$G_0' \simeq \frac{1}{m^2}\frac{1}{y^3}. \qquad (6.87)$$

We may estimate the value of y at which we expect (6.81) to break down. For $|\alpha| = 1$ we have, from the expressions for the derivatives, with $y \ll 1$

$$|\alpha(y)| = \frac{\pi^{\frac{3}{2}}}{2^{\frac{1}{2}}}\frac{1}{\lambda}\frac{e^{1/y}}{y^{\frac{1}{2}}} = 1. \qquad (6.88)$$

This gives the equation for the critical $y - y_0$; i.e.,

$$\ln(\lambda) - \ln\left(\frac{\pi^{\frac{3}{2}}}{2^{\frac{1}{2}}}\right) = \frac{1}{y_0} - \tfrac{1}{2}\ln(y_0). \qquad (6.89)$$

Thus (6.81) should be valid for

$$y > y_0 \simeq \frac{1}{\ln(\lambda)}. \qquad (6.90)$$

Table 6.1.

Particle	N^{DF}	Why?
e^+	2	Two spin states
e^-	2	Two spin states
ν_e	1	Left handed
$\bar{\nu}_e$	1	Right handed
ν_μ	1	Left handed
$\bar{\nu}_\mu$	1	Right handed
ν_τ	1	Left handed
$\bar{\nu}_\tau$	1	Right handed
γ	2	Two polarizations

It is instructive to take some specific examples to see what (6.90) means. We may begin with the original Lee–Weinberg problem (Lee and Weinberg, 1977) in which the L–$\bar{\mathrm{L}}$ pairs are taken to annihilate via the conventional weak interaction. For this case

$$\lambda = \frac{M_{\mathrm{pl}} m_{\mathrm{L}}}{(N^{\mathrm{DF}})^{\frac{1}{2}}} \frac{\langle \sigma v \rangle_{\mathrm{wk}}}{1.66}. \tag{6.91}$$

Anticipating later work we shall want this λ in the temperature range 10–100 MeV. We tabulate the degrees of freedom to be counted in this regime in Table 6.1. Thus, in this regime we expect that

$$N^{\mathrm{DF}} = 2 + (\tfrac{7}{8} \times 10) = 10.75.$$

The quantity $\langle \sigma v \rangle_{\mathrm{wk}}$ we take from the standard theory of weak interactions. Thus, ignoring possible distinctions among the rates in the various channels,

$$\langle \sigma v \rangle_{\mathrm{wk}} = N_{\mathrm{A}}(G_{\mathrm{F}}^2 m_{\mathrm{L}}^2 / 2\pi). \tag{6.92}$$

The factor N_{A} counts the open channels into which L, $\bar{\mathrm{L}}$ can annihilate. We have called these, generically, l, but more specifically for L particles of the mass we are anticipating they may be $\mathrm{L}\bar{\mathrm{L}} \to \nu_e\bar{\nu}_e, \nu_\mu\bar{\nu}_\mu, \nu_\tau\bar{\nu}_\tau, e^+e^-, \mu^-\mu^+, u\bar{u}$, $d\bar{d}$, and $s\bar{s}$. The latter three entries are quarks which we must count in triplicate for color. Thus

$$N_{\mathrm{A}} = 14.$$

The quantity G_F is the Fermi constant with

$$G_F \simeq 1.2 \times 10^{-5}/(\text{GeV})^2. \tag{6.93}$$

Hence, putting the numbers together we find

$$\lambda \simeq 5 \times 10^8 (m_L/\text{GeV})^3. \tag{6.94}$$

We will be dealing with $m_L \simeq$ GeV so that $\ln \lambda \simeq 20$ which, from (6.90) means that (6.81) is valid for $T \gtrsim m_L/20$. If we want to know N_L for $T \ll m_L/20$ we must resort to other techniques to which we return shortly.

The second example we shall study is e^+–e^- annihilation into photons, which occurs in a temperature regime appropriate to helium production, i.e., $T < 1$ MeV. The same number of degrees of freedom, N^{DF}, are still in equilibrium, at least at the beginning of annihilation. But here, where α stands for the fine structure constant,†

$$\langle \sigma v \rangle_{em} \simeq \frac{\alpha^2}{2\pi T^2} \simeq \frac{\alpha^2}{2\pi m_e^2}\,. \tag{6.95}$$

Thus

$$\lambda \simeq \frac{M_{pl}}{m_e}\frac{\alpha^2}{34} \simeq 3 \times 10^{16}, \tag{6.96}$$

which means $\ln \lambda \simeq 38$. Thus we expect in this case that (6.81) will be valid for $T > m_e/38$. Nonetheless, we can ask for $|\alpha(y)|$ for, say, $y = \frac{1}{10}$, which will bring us comfortably into the helium production regime. Using (6.88) and (6.96) we find $|\alpha(\frac{1}{10})| \simeq 10^{-11}$. This means one is entitled to use the equilibrium distribution for the electrons during this regime. The electrons also interact weakly with neutrinos, but this interaction, which is strongly temperature dependent, becomes negligible. These remarks constitute a justification for the usual treatment of helium production.‡ (See, for example, Weinberg (1972), where the equilibrium distributions for the leptons are used throughout.) Using the α-expansion we can write an expression for the rate of specific entropy generated during this regime. Thus, from (4.79), (4.60), and (6.41),

$$\dot{\sigma} = 2\alpha^2(y) n_{cl} \langle \sigma v \rangle. \tag{6.97}$$

† To avoid confusion we shall call the α we have been using $\alpha(y)$.

‡ For e^+–e^- annihilation the initial condition $\alpha(0) = 0$ is not quite right. There is a small electron excess. We return to this matter below.

We may compare this with the rate of expansion of the universe. Using (6.95) and (6.84) we find, for this example, that

$$\frac{\dot{\sigma}}{\dot{R}/R} \simeq 10^{-12},$$

which means that it is a very good approximation to take the entropy to be conserved.

As was mentioned earlier, once the condition given (6.90) is violated one must resort to more sophisticated methods to solve (6.73). The reason for wanting to solve this equation for small T, or y, is to discover how many L particles would, under various hypotheses, be around today. Since we know there is an ambient background of 3K microwave radiation consisting of some four hundred photons per cubic centimeter, it is convenient to use the photon density as a benchmark. This is easy to do by "renormalizing" G_0. To this end we note that, in equilibrium, the number of photons per volume is given by (recall the two polarization states)

$$n_\gamma = \frac{T^3}{\pi^2}\int_0^\infty \frac{\mathrm{d}x x^2}{\mathrm{e}^x - 1} = \frac{2T^3}{\pi^2}\xi(3), \tag{6.98}$$

where

$$\xi(3) = \textstyle\sum_{n=1}^{\infty}(1/n^3) \simeq 1.202. \tag{6.99}$$

Thus, we can define

$$\hat{G}(y) = n(y)/n_\gamma(y) \tag{6.100}$$

and

$$\hat{G}_0(y) = n_{\mathrm{cl}}(y)/n_\gamma(y), \tag{6.101}$$

with the asymptotic values

$$y \gg 1: \quad \hat{G}_0(y) = 1/\xi(3) \tag{6.102}$$

and

$$y \ll 1: \quad \hat{G}_0(y) \simeq \frac{1}{2\xi(3)}(\tfrac{1}{2}\pi)^{\frac{1}{2}}\frac{1}{\frac{1}{2}y^3}\mathrm{e}^{-1/y}, \tag{6.103}$$

leading to a modified rate equation

$$\frac{\mathrm{d}\hat{G}(y)}{\mathrm{d}y} = \hat{\lambda}[\hat{G}(y)^2 - \hat{G}_0(y)^2] \tag{6.104}$$

with

$$\hat{\lambda} = \frac{3\xi(3)}{\pi^3}\left(\frac{5}{\pi N^{\mathrm{DF}}}\right)^{\frac{1}{2}} m_{\mathrm{L}} M_{\mathrm{pl}} \langle \sigma v \rangle_{\mathrm{wk}} \simeq \tfrac{1}{4}\lambda. \tag{6.105}$$

We also have, as λ and $\hat{\lambda}$ differ by factors of $O(1)$, that $\hat{\lambda} \gg 1$. To deal with (6.104) we may linearize it by the transformation[†]

$$\hat{G}(y) = -\frac{1}{\hat{\lambda}}\frac{1}{f(y)}\frac{\mathrm{d}}{\mathrm{d}y} f(y). \tag{6.106}$$

Thus

$$\left(\frac{\mathrm{d}^2}{\mathrm{d}y^2} - \hat{\lambda}^2 \hat{G}_0(y)^2\right) f(y) = 0. \tag{6.107}$$

We can try a WKB solution to this zero energy "Schrödinger equation," with the same boundary condition as before. Thus during the regime in which the WKB is valid

$$f(y) \simeq \hat{G}_0(y)^{-\frac{1}{2}} \exp\left[-\hat{\lambda}\int^{y} \mathrm{d}y' G_0(y')\mathrm{d}y'\right] \tag{6.108}$$

or

$$\hat{G}(y) \simeq \hat{G}_0(y) + \frac{\hat{G}_0'(y)}{2\hat{\lambda}\hat{G}_0(y)^2}. \tag{6.109}$$

But this is just the expression we would find if we used the "renormalized" α of (6.81) in the suitably renormalized version of (6.74), which provides another way of looking at our previously derived approximate solution. With the slight change in constants the condition for the breakdown in the validity of (6.109) becomes

$$\frac{\hat{\lambda}}{\xi(3)}(\tfrac{1}{2}\pi y)^{\frac{1}{2}} \mathrm{e}^{-1/y} = 1, \tag{6.110}$$

which means the WKB will be valid, essentially so long as $y \gtrsim 1/\ln(\hat{\lambda})$.

For the regime $y < y_0$ we make the transformation

$$x = 1/y, \tag{6.111}$$

[†] This way of solving (6.104) was done in collaboration with L. Brown and G. Feinberg. For an independent treatment see Scherrer and Turner (1986).

and, in terms of a new function

$$f(y) = U(x)/x, \tag{6.112}$$

write (6.107) as

$$\left(\frac{\mathrm{d}^2}{\mathrm{d}x^2} - \frac{\pi\hat{\lambda}^2}{8[\xi(3)]^2}\frac{\mathrm{e}^{-2x}}{x}\right)U(x) = 0. \tag{6.113}$$

The object is now to use the **WKB** slightly above an x_0 defined by

$$\left(\frac{\pi\hat{\lambda}^2}{2[\xi(3)]^2}\right)^{\frac{1}{2}}\frac{\mathrm{e}^{-x_0}}{x_0^{\frac{1}{2}}} = 1, \tag{6.114}$$

and then to tie it smoothly to an approximate analytic solution to (6.113), which we are now going to proceed to develop. The **WKB** form for $U(x)$ can be obtained from (6.108). Thus

$$U(x) = \text{constant} \times x^{\frac{1}{4}}\exp\left[\tfrac{1}{2}x - \frac{\hat{\lambda}}{2\xi(3)}\left(\frac{\pi}{2x}\right)^{\frac{1}{2}}\mathrm{e}^{-x}\right]. \tag{6.115}$$

To extrapolate to $x \to \infty$ we make the replacement in (6.113)

$$\frac{\mathrm{e}^{-2x}}{x} \to \frac{\mathrm{e}^{-2x}}{x_0}.$$

We are taking advantage of the fact that most of the variation in this term, which is decreasing rapidly, is from the exponential. With this approximation (6.113) can be solved to yield

$$U(x) = K_0\left[\frac{\hat{\lambda}}{2\xi(3)}\left(\frac{\pi}{2x_0}\right)^{\frac{1}{2}}\mathrm{e}^{-x}\right], \tag{6.116}$$

where $K_0(x)$ is the modified Hankel function. Since, as the examples show, x_0 is a fairly large number, we can use the large argument form of the Hankel function and thus for x slightly below x_0 we have

$$U(x) \simeq \text{constant} \times x_0^{\frac{1}{4}}\exp\left[\tfrac{1}{2}x - \frac{\hat{\lambda}}{2\xi(3)}\left(\frac{\pi}{2x_0}\right)^{\frac{1}{2}}\mathrm{e}^{-x}\right], \tag{6.117}$$

which agrees, apart from the replacement of x by x_0 in two slowly varying factors, with (6.115). We can now reconstruct $\hat{G}$ in the large x, or small y, regime, which was the object of the exercise. Thus

$$\hat{G}(x) = \frac{1}{\hat{\lambda}}\left\{x^2\frac{\mathrm{d}}{\mathrm{d}x}\ln\left[K_0\left(\frac{\hat{\lambda}}{2\xi(3)}\left(\frac{\pi}{2x_0}\right)^{\frac{1}{2}}\mathrm{e}^{-x}\right)\right] - x\right\}. \tag{6.118}$$

Thus, for $x \gg 1$,

$$\hat{G}(x) = \frac{1}{\hat{\lambda}}\left\{\ln\left[\frac{\hat{\lambda}}{4\xi(3)}\left(\frac{\pi}{2x_0}\right)^{\frac{1}{2}}\right] + \gamma\right\}, \tag{6.119}$$

where $\gamma \simeq 0.577$ is Euler's constant. To the accuracy of our fitting procedure, x_0 is determined by (6.114).

Before we can apply $\hat{G}$ to the present epoch there is an important matter of principle involving e^+-e^- annihilation which we mush deal with. In the work done hitherto we have taken as an initial condition:

$$n_+(0) = n_-(0) = n_{cl}(0), \tag{6.120}$$

where $\pm$ refer to positron and electron, respectively. This condition is not quite correct for the real e^+-e^-. The point is that the universe is, as far as we know, electrically neutral and has no observable antiparticle component. This means there must be an electron excess since the electron charges are necessary to balance the total proton charge – as there does not appear to be antiproton charge available. Thus the correct initial condition in time is†

$$n_+(0)n_-(0) = n_{cl}^2(0). \tag{6.121}$$

To understand that this is correct, note that, neglecting quantum statistics, with $\alpha(t)$ the "effective" chemical potential, and $\mu(t)$ the "true" chemical potential,

$$n_+(t) = e^{-\mu(t)-\alpha(t)}n_{cl}(t), \tag{6.122}$$

while

$$n_-(t) = e^{+\mu(t)-\alpha(t)}n_{cl}(t). \tag{6.123}$$

Hence, assuming e^+ and e^- are in chemical and kinetic equilibrium initially, (6.121) is the correct initial condition. The rate equation for this process takes the form,‡

$$\frac{1}{R^3}\frac{d}{dt}(R^3 n_\pm) = \langle\sigma v\rangle(n_{cl}^2 - n_+ n_-). \tag{6.124}$$

† This formulation was done in collaboration with L. Brown.

‡ To see this begin with the equation

$$\frac{1}{R^3}\frac{\partial}{\partial t}[R^3 n_\pm(t)] = 4\iiiint (2\pi)^4\delta^{(4)}(p_+ + p_- - q - q')$$
$$\times W_1(p_+, p_-; q, q')\{[1 - f_+(p_+)][1 - f_-(p_-)]g(q)g(q')$$
$$- [1 + g(q)][1 + g(q')]f_+(p_+)f_-(p_-)\} \times dP_+\,dP_-\,dQ\,dQ',$$

(Footnote continues)

This leads to the conservation equation

$$R^3(n_- - n_+) = \text{constant}. \tag{6.125}$$

With the assumptions that lead to (6.73), we can write both of these equations in terms of the temperature; i.e.,

$$\frac{\mathrm{d}}{\mathrm{d}T}\left(\frac{n_+}{T^3}\right) = \frac{\langle\sigma v\rangle}{A^{\frac{1}{2}}}\left[\frac{n_+ n_-}{T^6} - \left(\frac{n(T)_{\text{cl}}}{T^3}\right)^2\right] \tag{6.126}$$

and

$$\frac{n_-}{T^3} - \frac{n_+}{T^3} = \text{constant}, \tag{6.127}$$

where A has the same meaning as before. Recalling, that, in equilibrium,

$$n_\gamma = 2\xi(3)T^3/\pi^2 \tag{6.128}$$

we have

$$\frac{\mathrm{d}}{\mathrm{d}T}\left(\frac{n_+}{n_\gamma}\right) = B\langle\sigma v\rangle\left[\frac{n_+ n_-}{n_\gamma^2} - \left(\frac{n_{\text{cl}}}{n_\gamma}\right)^2\right], \tag{6.129}$$

where

$$b = \frac{2\xi(3)/\pi^2}{A^{\frac{1}{2}}}, \tag{6.130}$$

while

$$\frac{n_-}{n_\gamma} - \frac{n_+}{n_\gamma} = \Delta, \tag{6.131}$$

where Δ is a constant whose value we shall now discuss.

We begin by noting that, putting in the factors of $\hbar$, c, k, the present photon density is given by

$$n_\gamma(T_p) = \frac{2.4}{\pi^2}\left(\frac{kT_p}{\hbar c}\right)^3. \tag{6.132}$$

(Footnote continued)

where $f_\pm$ are the Boltzmann functions for the leptons and g is the photon function. The respective quantum statistics have been indicated. If we now make the same set of approximations that led to (6.43) we may derive (6.124).

If we take $T_p = 2.7\text{K}$ we find

$$n_\gamma(T_p) \simeq 398/\text{cm}^3. \tag{6.133}$$

The age of the universe, t_p, is about 1.5×10^{10} years, or about 5×10^{17} s. The size of the visible universe is about 1.5×10^{28} cm, so the volume is about 3.4×10^{84} cm^3. Thus the total number of photons in the visible universe is at present,

$$N_\gamma(T_p) \simeq 10^{87}.$$

The number of baryons is less certain but the usual estimate is

$$N_B/N_\gamma \simeq 10^{-9} \text{ to } 10^{-10}. \tag{6.134}$$

We shall see that after annihilation the number of positrons is sensibly zero. Hence to balance the proton charge we must have

$$N_{e^-}/N_\gamma \simeq 10^{-9} \text{ to } 10^{-10}. \tag{6.135}$$

Thus, from (6.131)

$$\Delta \simeq 10^{-9} \text{ to } 10^{-10}. \tag{6.136}$$

It is true, as far as we know, that at present $n_- \gg n_+$. We are going to solve for n_+ using this condition, extrapolated back to near the beginning of the annihilation. We begin with the equation,

$$\frac{\mathrm{d}}{\mathrm{d}T}\left(\frac{n_+}{n_\gamma}\right) = B\langle\sigma v\rangle\left[\frac{n_+ n_-}{n_\gamma^2} - \left(\frac{n_{\text{cl}}}{n_\gamma}\right)^2\right]. \tag{6.137}$$

Using the relative magnitudes of n_+ and n_- in (6.131) we will solve this equation with the approximate condition that

$$\frac{n_-}{n_\gamma} \simeq \Delta. \tag{6.138}$$

Thus we are to solve the approximate linear equation

$$\frac{\mathrm{d}}{\mathrm{d}T}\left(\frac{n_+}{n_\gamma}\right) = B\langle\sigma v\rangle\left[\Delta\frac{n_+}{n_\gamma} - \left(\frac{n_{\text{cl}}}{n_\gamma}\right)^2\right], \tag{6.139}$$

subject to the initial condition in time,

$$\frac{n_+(0)n_-(0)}{n_\gamma^2(0)} = \frac{n^2{}_{\text{cl}}(0)}{n_\gamma^2(0)}. \tag{6.140}$$

Once again, letting $Y = T/m_e$ and,

$$\hat{G}(y) = \frac{n_+(y)}{n_\gamma(y)}, \tag{6.141}$$

we have the equation to solve,

$$\frac{d}{dy}(\hat{G}) = \hat{\lambda}(\Delta\hat{G} - \hat{G}_0^2), \tag{6.142}$$

with $\hat{\lambda}$, which is in general a function of y given by

$$\hat{\lambda} = m_e B\langle\sigma v\rangle \tag{6.143}$$

and

$$\hat{G}_0 = \frac{n_{cl}}{n_\gamma}. \tag{6.144}$$

The boundary condition is now†

$$\hat{G}(0) = \hat{G}_0^2(0)\frac{1}{\Delta}. \tag{6.145}$$

It is convenient to write the solution of (6.142) in the form

$$\hat{G}(y) = \hat{G}_0^2(y)\frac{1}{\Delta} + \bar{G}(y), \tag{6.146}$$

where $\bar{G}$ obeys the boundary condition,

$$\bar{G}(0) = 0. \tag{6.147}$$

Thus $\bar{G}$ obeys the equation

$$\frac{d}{dy}\left(\frac{\hat{G}_0^2(y)}{\Delta}\right) + \frac{d\bar{G}(y)}{dy} = \hat{\lambda}\,\Delta\bar{G}(y) \tag{6.148}$$

so that, in light of (6.147)

$$\begin{aligned}\bar{G}(y) &= -\int_{y_0}^{y}\exp\left(-\int_{y_0}^{y'}\Delta\hat{\lambda}\,dy''\right)\frac{d}{dy'}\left(\frac{\hat{G}_0^2(y')}{\Delta}\right)dy'\exp\int_{y_0}^{y}\Delta\hat{\lambda}\,dy' \\ &= -\int_{y_0}^{y}\exp\left(\int_{y'}^{y}\Delta\hat{\lambda}\,dy''\right)\frac{d}{dy'}\left(\frac{\hat{G}_0^2(y')}{\Delta}\right)dy' \\ &= \int_{y}^{y_0}\exp\left(-\int_{y}^{y'}\Delta\hat{\lambda}\,dy''\right)\frac{d}{dy'}\left(\frac{\hat{G}_0^2(y')}{\Delta}\right)dy',\end{aligned} \tag{6.149}$$

† The $y = 0$ limit corresponds to $t = \infty$.

where we have changed the signs in the last equation to reflect the fact that $y < y_0$. One use we can make of these equations is to deduce how many positrons are present in the limit as $y \to 0$. To this end we define y_f, i.e., y "freezing," as follows:

$$\int_0^{y_f} \Delta\hat{\lambda}\,dy' = 1. \tag{6.150}$$

Since $\hat{\lambda}$ is a monotone function of y in the examples we shall study, we have

$$\begin{aligned} \int_0^{y} \Delta\hat{\lambda}\,dy' &> 1, \quad y > y_f \\ &< 1, \quad y < y_f. \end{aligned} \tag{6.151}$$

Hence, in the integral occurring in (6.149), we shall replace $\exp(-\int_0^{y'} \Delta\hat{\lambda}\,dy'')$ by

$$\begin{aligned} \theta(y_f) &= 1, \quad y < y_f \\ &= 0, \quad y > y_f. \end{aligned} \tag{6.152}$$

With this approximation,

$$\bar{G}(y) = \int_0^{y_f} \frac{d}{dy'}\left(\frac{\hat{G}_0^2(y')}{\Delta}\right)dy' = \frac{\hat{G}_0^2(y_f) - \hat{G}_0^2(0)}{\Delta}. \tag{6.153}$$

Thus we have, approximately,

$$\hat{G}(0) = \frac{\hat{G}_0^2(y_f)}{\Delta}. \tag{6.154}$$

To compute y_f we write, collecting definitions,

$$\hat{\lambda} = \frac{\xi(3)}{\pi^3}\alpha^2 \frac{1}{(8\pi^3 N^{DF}/90)^{\frac{1}{2}}} \frac{M_{pl}}{m_e} \simeq 10^{16} \tag{6.155}$$

where we have used the same estimate of N^{DF} as above. Thus, taking $\Delta \simeq 10^{-10}$ and using (6.150) we have† $y_f \simeq 10^{-6}$. We may use this y_f to evaluate $\hat{G}(0)$. We have

$$\hat{G}(y_f) \simeq \frac{1}{2\xi(3)}\left(\tfrac{1}{2}\pi\right)^{\frac{1}{2}} \frac{1}{y_f^{\frac{3}{2}}} e^{-1/y_f}. \tag{6.156}$$

† This calculation has assumed that the universe is "radiation" dominated; that ρ is dominated by massless particles. For y as small as y_f this is unlikely to be true. In this regime ρ will be dominated by massive baryons. Making a crude

(Footnote continues)

Therefore

$$\hat{G}(0) \simeq 10^{10} \times 10^{18} \times e^{-2\times 10^6} \simeq 0.$$

Thus one should not expect to find any measurable number of relic positrons. We also want to estimate $\hat{G}(y)$ for $y \gg y_f$. For this region we can use the equilibrium condition $\alpha \simeq 0$ since annihilations are compensated for by recombination. From (6.122), (6.123), and (6.131) we have the relation,

$$4/\hat{G}_0 = e^{\mu-\alpha} - e^{-\mu-\alpha} \simeq 2\sinh\mu \simeq 2\mu. \tag{6.157}$$

The validity of this last step rests on the smallness of μ which, using the definition of $\hat{G}_0$, will be true, essentially, so long as

$$y > -\frac{1}{\log\Delta}. \tag{6.158}$$

In the case at hand this means so long as $y > \frac{1}{20}$. It is interesting to get a feeling for how rapidly the positrons are disappearing as a function of y in this regime. For this purpose, and others, the table given below is useful.†
The quantity we present is

$$\hat{G}_0(y) = \frac{1}{2\xi(3)}\int_0^\infty x^2 \frac{1}{\exp(x^2 + 1/y^2)^{\frac{1}{2}} + 1}\,dx. \tag{6.159}$$

The inclusion of the correct quantum statistical factor makes a difference of a few percent for $y > 1$. We give $\hat{G}_0(y)$ for fifty values ranging from ∞ to 2×10^{-2}; see Table 6.2.

(Footnote continued)

estimate of how this changes the results, and ignoring all $O(1)$ constants and calling M the mass of the proton, we have

$$-\frac{\dot{T}}{T} = \frac{M^{\frac{1}{2}}}{M_{\rm pl}} T^{\frac{3}{2}}.$$

Hence, the freezing condition becomes approximately

$$\Delta\alpha^2 \frac{M_{\rm pl}}{m_e}\left(\frac{m_e}{M}\right)^{\frac{1}{2}} y_f^{\frac{3}{2}} = 1,$$

which gives

$$y_f \simeq 10^{-4}.$$

This does not change the conclusions.

† Table 6.2 was prepared in collaboration with G. Feinberg.

Table 6.2. *The* $\hat{G}_0(y)$ *table*

y	$\hat{G}_0(y)$	$1/y$	y	$\hat{G}_0(y)$	$1/y$
∞	0.75	0	3.9×10^{-2}	3.8×10^{-10}	26
1	0.63	1	3.7×10^{-2}	1.5×10^{-10}	27
0.5	0.41	2	3.6×10^{-2}	5.7×10^{-11}	28
0.33	0.23	3	3.5×10^{-2}	2.2×10^{-11}	29
0.25	0.12	4	3.3×10^{-2}	8.5×10^{-12}	30
0.20	5.4×10^{-2}	5	3.2×10^{-2}	3.3×10^{-12}	31
0.17	2.5×10^{-2}	6	3.1×10^{-2}	1.3×10^{-12}	32
0.14	1.1×10^{-2}	7	3.0×10^{-2}	4.9×10^{-13}	33
0.125	5.0×10^{-3}	8	2.9×10^{-2}	1.9×10^{-13}	34
0.11	2.1×10^{-3}	9	2.85×10^{-2}	7.2×10^{-14}	35
0.10	9.2×10^{-4}	10	2.8×10^{-2}	2.8×10^{-14}	36
9.1×10^{-2}	3.8×10^{-4}	11	2.7×10^{-2}	1.1×10^{-14}	37
8.3×10^{-2}	1.6×10^{-4}	12	2.6×10^{-2}	4.0×10^{-15}	38
7.7×10^{-2}	6.3×10^{-5}	13	2.56×10^{-2}	1.5×10^{-15}	39
7.1×10^{-2}	2.6×10^{-5}	14	2.5×10^{-2}	5.9×10^{-16}	40
6.7×10^{-2}	1.0×10^{-5}	15	2.44×10^{-2}	2.2×10^{-16}	41
6.3×10^{-2}	4.2×10^{-6}	16	2.38×10^{-2}	8.5×10^{-17}	42
5.9×10^{-2}	1.7×10^{-6}	17	2.33×10^{-2}	3.3×10^{-17}	43
5.6×10^{-2}	6.7×10^{-7}	18	2.3×10^{-2}	1.2×10^{-17}	44
5.3×10^{-2}	2.7×10^{-7}	19	2.2×10^{-2}	4.7×10^{-18}	45
5.0×10^{-2}	1.1×10^{-7}	20	2.17×10^{-2}	1.8×10^{-18}	46
4.8×10^{-2}	4.2×10^{-8}	21	2.1×10^{-2}	6.8×10^{-19}	47
4.6×10^{-2}	1.6×10^{-8}	22	2.08×10^{-2}	2.6×10^{-19}	48
4.4×10^{-2}	6.4×10^{-9}	23	2.04×10^{-2}	9.8×10^{-20}	49
4.2×10^{-2}	2.5×10^{-9}	24	2.00×10^{-2}	3.7×10^{-20}	50
4.0×10^{-2}	9.8×10^{-10}	25			

In the standard treatments of e^+–e^- annihilation (for example Weinberg, 1972) it is assumed for $T < m_e \simeq 5 \times 10^9$ K that the electrons and positrons have "disappeared," i.e., that they make no significant contribution to the energy density or entropy of the universe. What one means by "significant" or "disappeared" is, to some extent, in the eyes of the beholder. Table 6.2, which should be reliable during the epoch of helium formation, makes clear how the relative electron–positron abundance decreases as a function of y.

Since both the effective chemical potential α and the actual chemical potential μ are sensibly zero during this epoch we know from our previous discussion that the entropy current of the electrons, positrons, and photons is approximately conserved. There is a second entropy current due to neutrinos which, we will now argue, is separately conserved. The point is, that prior to this epoch, neutrinos have decoupled from the electrons and hence the photons. To see this we compare $\dot{R}/R$ and the neutrino collison rate. Ignoring all factors of $O(1)$

$$\frac{\dot{R}}{R} \simeq \frac{T^2}{M_{\text{pl}}}, \tag{6.160}$$

while

$$n\langle\sigma v\rangle_{\text{wk}} \simeq T^3 \times G_{\text{F}}^2 T^2. \tag{6.161}$$

If we equate the two rates we get the condition

$$T_{\text{c}}^3 = \frac{1}{M_{\text{pl}} G_{\text{F}}^2} \simeq 10^{10} \frac{\text{GeV}^4}{M_{\text{pl}}} \simeq 10^{-9}\ \text{GeV}^3. \tag{6.162}$$

Thus

$$T_{\text{c}} \simeq 1\ \text{MeV},$$

which agrees, essentially, with the more careful estimates. T_{c} is the temperature at which a neutrino would have a single collision with an electron during an e-folding time. This is the criterion we shall use for decoupling. Hence for $T < T_{\text{c}}$ the neutrinos expand freely preserving their equilibrium Boltzmann distributions.†

We have argued (see Appendix B for further discussion) that for a gas described by approximate equilibrium distributions the total entropy $R^3 s$ is given by, for zero chemical potential,

$$S = \frac{U + P_0 R^3}{T}, \tag{6.163}$$

where U is the total energy, $R^3\rho$, and P_0 is the pressure computed by using the equilibrium distribution. For massless particles in general

$$PR^3 = \tfrac{1}{3}U. \tag{6.164}$$

Thus the neutrino component of the entropy, which is separately conserved

† Different "flavors" of neutrinos – ν_e, ν_μ, ν_τ – will decouple at somewhat different T_{c}s, but that is a nuance which need not concern us here.

after decoupling, is given by

$$S_\nu = \frac{4}{3}\frac{U_\nu}{T^\nu} = \frac{7}{90}\pi^2(RT^\nu)^3 N_F, \tag{6.165}$$

where N_F is the number of neutrino flavors. If only the ν_e, ν_μ, and ν_τ exist then $N_F = 3$. From the conservation of S_ν it follows that

$$RT^\nu = \text{constant}. \tag{6.166}$$

We may now turn our attention to the e^+, e^-, γ sector with its separately (approximately) conserved entropy s. Using (6.163) we may write

$$S = \frac{\frac{4}{3}U_\gamma + U_1 + P_1}{T}. \tag{6.167}$$

The subscript "l" refers to both electrons and positrons and T is the electron–positron γ temperature which, as we shall see, is different from T^ν. It is instructive to give S in the two limits: $T > m$ and $T < m$. For $T \gg m$, using (6.64)

$$S_> \simeq \frac{4}{3}\frac{(U_\gamma + U_e)}{T} = \frac{22}{90}\pi^2(RT)^3. \tag{6.168}$$

For $T \ll m$, using classical statistics for e^+, e^- we have, using (4.31) and (6.98)

$$S_< \simeq \frac{4\pi^2}{45}(RT)^3\left[1 + \left(\frac{m}{T} + \frac{5}{2}\right)\left(\frac{n_+}{n_\gamma} + \frac{n_-}{n_\gamma}\right)\frac{90}{19}\right]. \tag{6.169}$$

We can now choose a temperature, $T_<$, such that the electron-positron densities are vanishingly small, i.e., such that

$$S_< \simeq \frac{8\pi^2}{90}(R_< T_<)^3 \tag{6.170}$$

and let us choose a temperature $T_>$ such that S is given by (6.168). Thus

$$S_> \simeq \frac{22}{90}\pi^2(R_> T_>)^3. \tag{6.171}$$

From the conservation of S we have

$$S_> \simeq \frac{22}{90}\pi^2(R_> T_>)^3 = S_< \simeq \frac{8}{90}\pi^2(R_< T_<)^3. \tag{6.172}$$

Thus

$$\frac{T_<}{T_>} = \left(\frac{11}{4}\right)^{\frac{1}{3}} \frac{R_>}{R_<} = \left(\frac{11}{4}\right)^{\frac{1}{3}} \frac{T^{\nu}_<}{T^{\nu}_>}, \tag{6.173}$$

where we have used (6.166). If we extrapolate this formula to high enough temperatures so that

$$T_> \simeq T^{\nu}_> \tag{6.174}$$

we learn, dropping the subscripts, that after $e^- - e^+$ annihilation

$$T = \left(\frac{11}{4}\right)^{\frac{1}{3}} T^{\nu}, \tag{6.175}$$

a formula that should hold at present assuming the neutrinos ν_e, ν_μ, and ν_τ are massless.†

We may now apply these considerations to determining the present density of the putative L particles. We shall argue that, in order not to overclose the present universe, the masses of such particles must exceed 1 GeV. Assuming they have weak interactions, as we have been doing, they will decouple from e^+, e^-, ν at about 1 MeV. During much of this regime (6.119) will apply. Since $\hat{\lambda} \gg 1$ we can write, using (6.114),

$$\begin{aligned} n(x \gg 1) &= n_\gamma(x \gg 1)\hat{G}(x \gg 1) \\ &\simeq n_\gamma(x \gg 1)\frac{1}{\hat{\lambda}} \ln\left(\frac{\hat{\lambda}}{(\ln \hat{\lambda})^{\frac{1}{2}} 4}\right). \end{aligned} \tag{6.176}$$

This formula will hold only prior to $e^+ - e^-$ annihilation. After that, annihilation is essentially completed, so that (6.175) obtains; i.e., the photons will have suffered a temperature increase of $(\frac{11}{4})^{\frac{1}{3}}$ compared to the freely expanding L particles. Hence the photon density will have increased by a relative factor of $\frac{11}{4}$. Thus to apply (6.176) to the present, we must rescale. Hence at present

$$n(T_p) \simeq \frac{4}{11} n_\gamma(T_p) \frac{1}{\hat{\lambda}} \ln\left(\frac{\hat{\lambda}}{(\ln \hat{\lambda})^{\frac{1}{2}} 4}\right). \tag{6.177}$$

How large an $n(T_p)$ can we stand? We do not want to overclose the universe which, with its present energy density, appears to be nearly flat. The visible baryon mass density, ρ_B, is about, with the uncertainties indicated,

$$\rho_B = 0.02 - 0.20 \times 10^{-29}\ \text{g/cm}^3 < 10^{-3}\ \text{MeV/cm}^3. \tag{6.178}$$

† For a discussion of the massive case see Appendix C.

We may compare this to the so-called critical density, ρ_c, defined so that the equation

$$\frac{\dot{R}^2}{R^2} = \frac{8\pi}{3}\frac{1}{M_{pl}^2}\rho_c \tag{6.179}$$

is valid. This is the density which, if one were to know the present value of $\dot{R}/R$, would allow the universe to be flat. However, $\dot{R}/R$ is uncertain to within a factor of two. In fact calling, as is customary, the present value of $\dot{R}/R$, H_0 – the "Hubble constant" – we have from observation, roughly,

$$50\,\mathrm{km/s\,M_{pc}} < H_0 < 100\,\mathrm{km/s\,M_{pc}}, \tag{6.180}$$

where to conform to convention we have introduced the customary astronomer's unit, $\mathrm{M_{pc}}$, for "megaparsec."

$$\begin{aligned} 1\ \text{parsec} &= 3.09 \times 10^{13}\,\mathrm{km} \\ &= 3.26\ \text{light years.} \end{aligned} \tag{6.181}$$

Thus the unit

$$\frac{\mathrm{km}}{\mathrm{s\,M_{pc}}} = \frac{1}{3.09 \times 10^{19}\,\mathrm{s}} = \frac{1}{9.78 \times 10^{11}\ \text{years}}. \tag{6.182}$$

Thus

$$\begin{aligned} \rho_c &= H^2 M_{pl}^2 \frac{3}{8\pi} \\ &= \left(\frac{H}{75\,\mathrm{km/s\,M_{pc}}}\right)^2 \times 1.05 \times 10^{-29}\,\mathrm{g/cm^3} \\ &= \left(\frac{H}{75\,\mathrm{km/s\,M_{pc}}}\right)^2 \times 5.6 \times 10^{-3}\,\mathrm{MeV/cm^3}. \end{aligned} \tag{6.183}$$

On the same scale the energy density of the radiation at present, ρ_γ, is given by

$$\rho_\gamma \simeq 3 \times 10^{-7}\,\mathrm{MeV/cm^3}. \tag{6.184}$$

From (6.94) we see that

$$\hat{\lambda} \sim m_L^3.$$

Hence if at some mass, m_L, the present heavy lepton energy density counting both L and $\bar{\mathrm{L}}$, which is given by

$$\rho_L(T_p) = 2m_L n(T_p) \simeq \tfrac{8}{11} n_\gamma(T_p)\frac{1}{\hat{\lambda}}\ln\left(\frac{\hat{\lambda}}{4\ln(\hat{\lambda})^{\frac{1}{2}}}\right)m_L, \tag{6.185}$$

is just equal to, say, ρ_c then for masses greater[†] than m_L, ρ_L will be less than ρ_c. In other words, the requirement that we do not overclose the universe sets a *lower* bound on m_L, a perhaps surprising result that was first noted by Lee and Weinberg (1977). Putting in the numbers gives[‡]

$$m_L \gtrsim 2 \text{ GeV}.$$

Implicit in this result is the expression for $\langle\sigma v\rangle_{wk}$ given by (6.92). The form of this expression arises from a Feynman diagram in which the $L\bar{L}$ annihilate into a virtual Z^0 whose on-shell rest mass is given by

$$m_{Z^0} = 92.6 \pm 1.7 \text{ GeV}.$$

The Z^0, in turn, interacts with the $\nu\bar{\nu}$ pairs, and the like. The matrix element responsible for (6.92) is, up to inessential factors, given by

$$M_{wk} \simeq \left(\frac{g}{2\cos\theta_w}\right)^2 \frac{1}{q^2 + m_Z{}^2}, \tag{6.186}$$

where g is the electroweak coupling constant and θ_w the Weinberg angle. The regime we have been considering is one in which $|q^2|/m_Z^2 \ll 1$, where $|q^2| > 4m_L^2$.* But if we allow m_L to become very large then the q^2 term will dominate the propagator in (6.186). The effect of this is, approximately, to replace the $\hat{\lambda}$ of (6.105) by

$$\hat{\lambda}' = \left(\frac{m_Z}{m_L}\right)^4 \hat{\lambda}. \tag{6.187}$$

Using this modified $\hat{\lambda}'$ in (6.185) we see that, apart from logarithmic factors, in this regime $\rho_L(T_p)$ *increases* as the square of m_L. Hence there is an *upper bound* to masses of the L we can tolerate which is, putting in the numbers,

$$m_L < 10^5 \text{ GeV}.$$

The regime of very light, but still massive, neutrinos, we discuss in Appendix C. We next turn to the matter of unstable particles.

† As we shall shortly argue, for $m \gg m_L$ this argument will break down.

‡ A very careful analysis of this number can be found in E. W. Kolb and K. A. Olive (1986).

* We are using the +, +, +, − signature.

7
Unstable particles

The burden of Chapter 6, taken with Appendix C, was to place mass limits on stable neutral leptons – which we shall refer to generically as "neutrinos." By "stable" we mean neutrinos whose lifetimes are greater than 10^{17} s – the order of magnitude of the present age of the universe. These mass limits resulted from the requirement that the neutrinos do not overclose the universe. For "heavy" neutrinos these arguments led to the statement that we must have

$$m_{\mathrm{L}} > 2 \text{ GeV},$$

while for "light" neutrinos

$$\sum_{\text{flavors}} m_i < 50 \text{ eV},$$

with the understanding that these limits can be changed, by perhaps a factor of two, by modifying some of the assumptions. At about the same time these arguments were first made, it was realized (Dicus, Kolb, and Teplitz, 1977) that the limits could be completely changed if the neutrinos were unstable. By now, an immense literature has accumulated on unstable neutrinos.† The reason for this literature explosion is that theorists realized that many different kinds of bounds could be set depending on how these neutrinos are presumed to decay. It is not our purpose to review this literature, which would require a book in itself, but to focus on some of the statistical mechanical issues. However, we will at least give some of the flavor of these questions.

Current ideas, in particular the apparent quark–lepton analogy – the fact that quark flavors and lepton flavors appear to go hand in hand, with

† A very useful summary with innumerable references can be found in Sarkar (1986) or in the articles in Fackler and Thanh Van (1986). A recent, and detailed, review is given in Harari and Nir (1987).

"nonstrange" quarks having the lightest mass – suggest that

$$m_{\nu_e} < m_{\nu_\mu} < m_{\nu_\tau}.^\dagger$$

It is, one should bear in mind, quite possible that these masses are all actually zero. If these neutrinos do have a mass then it is certainly conceivable that some of them are unstable. Since ν_e is presumed to be the lightest lepton it is plausible to suppose that it is essentially stable. Some of the more conventional decay modes for unstable neutrinos which have been considered are, using numerical subscripts for the various kinds of neutrinos,

$$\nu_2 \rightarrow \nu_1 + \gamma$$

$$\nu_2 \rightarrow \nu_1 + \gamma + \gamma$$

or

$$\nu_2 \rightarrow \nu_1 + \nu_1 + \bar{\nu}_1,$$

as well as, with $m_{\nu_2} > 2m_e$,

$$\nu_2 \rightarrow e^+ + e^- + \nu_1.$$

In addition, all sorts of decays into unconventional particles such as "familons" or "majorons" have been proposed and studied theoretically. One is, however, struck by how little is actually known empirically about the light neutrinos. (Of the heavy neutrinos, nothing is known experimentally, because none have been observed.) We have already given the relatively poor mass limits on these neutrinos. The lifetime limits are even worse. The best lifetime limit is on the ν_e – actually the $\bar{\nu}_e$ – and it comes from electron neutrinos and especially antineutrinos, which are easier to detect, that arrived on Earth on February 23, 1987 from the supernova SN1987A in the Large Magellanic Cloud, visible optically in the southern hemisphere. Astronomers use the unit "kiloparsec" to measure such huge distances where (see (6.181))

$$1 \text{ kpc} = 3.09 \times 10^{21} \text{ cm}.$$

The mean distance of the Large Magellanic Cloud is about 50 kpc. These neutrinos (antineutrinos) had an average energy of about 10 MeV. We

† The neutrinos, if they are massive, will in general be "mixed"; i.e., the states with definite flavor do not have definite mass or, conversely, the mass states that propagate do not have a definite flavor. It is customary to call these mass states ν_1, ν_2, and ν_3. We have used a more suggestive, if less precise, notation.

know from terrestrial experiments that $m_{\nu_e} < 10$ eV. Hence the electron neutrinos arriving from the supernova are highly relativistic with $v \simeq c$. Thus, the time of flight, t, from the supernova is given by

$$t \simeq 5 \times 10^{12}\ \text{s}.$$

The proper time, t_0, is related to t by the equation

$$t_0 = \frac{m}{E}\, t. \tag{7.1}$$

If we take $m \simeq 10$ eV and $E \simeq 10$ MeV

$$t_0 \simeq 5 \times 10^6\ \text{s}.$$

We can then roughly state that the supernova sets a lower limit on the lifetime given by

$$t_{\nu_e} > 1\ \text{year}.$$

This shows how poorly this lifetime is known directly from experiment. The other lifetimes are even less well known. For example, if $m_\tau > 2m_e$, the decay

$$\nu_\tau \to e^+ + e^- + \nu_e$$

is energetically possible. The experiment of Bergsma *et al.* (1983) yields a limit of something like $t_{\nu_\tau} > 200-1000$ s for $m_{\nu_\tau} \simeq 30-200$ MeV. The muon neutrino's lifetime is still less well known.

Perhaps the most straightforward attempt to set cosmological limits on masses and lifetimes of unstable neutrinos is to repeat the discussion of the last chapter using the upper bound on the energy density, but now allowing for instability of the neutrinos. This program was initiated in the paper of Dicus, Kolb, and Teplitz (1977). For definiteness we shall consider the putative decay mode

$$\nu_L \to \nu_l + \nu_l + \bar{\nu}_l.$$

As in the last chapter, we shall call $n(t)$ the number density distribution of the L particles as a function of time. We shall call $f(p)$ the Boltzmann distribution of the L particles and $g(p)$ the Boltzmann distribution of the l and $\bar{l}$ particles. We will now show that n obeys the equation, including $L\bar{L}$ annihilation, under a set of assumptions to be specified,

$$\frac{1}{R^3}\frac{\partial}{\partial t}(R^3 n) = -\langle \sigma v \rangle (n^2 - n^2_{\text{cl}}) - \langle \Gamma \rangle (n - n_{\text{cl}}). \tag{7.2}$$

The notation is that of the last chapter with the exception of $\langle\Gamma\rangle$ which is to be defined. The first term on the right-hand side of (7.2) is now familiar. It is the second term, which represents decay of the L, that is new. Before making any simplifying assumptions the decay term takes the form

$$2\int\frac{d^3p}{(2\pi)^3E(p)}\int\frac{d^3q_1}{(2\pi)^32E(q_1)}\int\frac{d^3q_2}{(2\pi)^32E(q_2)}\int\frac{d^3\bar{q}_2}{(2\pi)^32E(\bar{q}_2)}W_D(p;q_1,q_2,\bar{q}_2)$$
$$\times(2\pi)^4\delta^{(4)}(p-q_1-q_2-\bar{q}_2)\{g(q_1)g(q_2)g(\bar{q}_2)[1-f(p)]$$
$$-f(p)[1-g(q_1)][1-g(q_2)][1-g(\bar{q}_2)]\}. \tag{7.3}$$

We may now make the same set of assumptions as in the last chapter; i.e.,

(a) The l-particles thermalize rapidly so that g can be replaced by a Fermi–Dirac distribution.
(b) The $f(p)$ is characterized by a pseudochemical potential $\alpha(t)$; i.e., $f\simeq e^{-\alpha(t)-\beta E_L}$.
(c) The L-particles are nonrelativistic when they decay.

With these assumptions (7.3) can be written as

$$2\int\frac{d^3p}{(2\pi)^3m_L}\int\frac{d^3q_1}{(2\pi)^32q_1}\int\frac{d^3q_2}{(2\pi)^32q_2}\int\frac{d^3\bar{q}_2}{(2\pi)^32\bar{q}_2} \tag{7.4}$$
$$\times(2\pi)^4\delta(m_L-q_1-q_2-\bar{q}_2)\delta^{(3)}(\mathbf{q}_1+\mathbf{q}_2+\bar{\mathbf{q}}_2)$$
$$\times W_D(p;q_1,q_2,\bar{q}_2)(1-e^{-\alpha})\{f_0(p)$$
$$\times[1-g_0(q_1)][1-g_0(q_2)][1-g_0(\bar{q}_2)]\}$$
$$\equiv\langle\Gamma\rangle(n_{cl}-n),$$

where $\langle\Gamma\rangle$ is given by

$$\langle\Gamma\rangle=\frac{1}{m_L}\int\frac{d^3q_1}{(2\pi)^32q_1}\int\frac{d^3q_2}{(2\pi)^32q_2}\int\frac{d^3\bar{q}_2}{(2\pi)^32\bar{q}_2} \tag{7.5}$$
$$\times W_D(p;q_1,q_2,\bar{q}_2)(2\pi)^4\delta(m_L-q_1-q_2-\bar{q}_2)$$
$$\times\delta^{(3)}(\mathbf{q}_1+\mathbf{q}_2+\bar{\mathbf{q}}_2).$$

In general, one would solve (7.2), find n in the limit $t\to\infty$, and repeat the discussion of the last chapter. However, the usual scenario that is explored is to suppose that the temperature at which the decays take place, T_d, is much *lower* than the temperature T_f at which the number of L particles freeze out after annihilation. Or, in terms of the corresponding times $t_d \gg t_f$. With this scenario, the problem divides conveniently into two parts. First we can use the results of the last chapter to determine $n(T_f)$ and then secondly use this as an initial condition in solving the decay problem. This is the way the problem was done in the literature (Dicus, Kolb, and Teplitz, 1978). However, these authors equate $n(T_f)$ with $n_{cl}(T_f)$, the equilibrium distribution. We have seen in the last chapter that this is not, in general, a very good approximation, and we shall not make it.† In fact, for $t > t_f$, because of its exponentially decreasing character, we have, $n_{cl}(t) \ll n(t)$ and hence during this regime (7.2) becomes approximately,

$$\frac{\partial}{\partial t}(R^3 n) = -\langle\Gamma\rangle R^3 n. \tag{7.6}$$

which has, taking into account the initial condition, and supposing that $\langle\Gamma\rangle$, defined by (7.5), is a slowly varying function of t, the solution

$$n(t) = n(t_f)\left(\frac{R(t_f)}{R(t)}\right)^3 e^{-\Gamma(t-t_f)}, \tag{7.7}$$

which has the obvious limiting property that when $\Gamma = 0$,

$$R^3 n = \text{constant}. \tag{7.8}$$

We would like to know the present value of the energy density, ρ_p, due to these decaying particles and their decay products. The contribution to ρ_p due to the residual L-particles, ρ_{Lp}, can be found directly from (7.7); i.e.,

$$\rho_{Lp} = m_L n(t_f)\left(\frac{R(t_f)}{R(t_p)}\right)^3 e^{-\Gamma(t_p-t_f)}. \tag{7.9}$$

Because of the rapidly decreasing exponential this is not expected to be the significant contribution to the energy density. This will come rather from ρ_l, the energy density of the light neutrinos into which the L particles decay. These neutrinos carry an average initial energy of $m_L/3$ apiece, which becomes redshifted as the universe expands. The energy density of any of these light neutrinos is given by

$$\rho_\nu = \int \frac{d^3 q}{(2\pi)^3}\, q g(q,t), \tag{7.10}$$

where $g(q,t)$ obeys, in this regime, the equation‡

$$\frac{\partial g}{\partial t} - \frac{\dot{R}}{R} q \frac{\partial g}{\partial q} = \frac{1}{q} C_D(q,t), \tag{7.11}$$

† The work to be described here was done in collaboration with L. Brown.

‡ g is the Boltzmann function for a single spin degree of freedom.

where

$$C_D = -\iiint (2\pi)^4 \delta^{(4)}(p' - q - q' - \bar{q}')$$
$$\times W_D(p'; q, q', \bar{q}')\{g(q)g(q')g(\bar{q}')[1 - f(p')]$$
$$- [1 - g(q)][1 - g(q')][1 - g(\bar{q}')]f(p')\}\, dQ'\, d\bar{Q}'\, dP'. \quad (7.12)$$

Thus

$$\frac{\partial \rho_\nu}{\partial t} = \int \frac{d^3 q}{(2\pi)^3}\, q\, \frac{\partial g}{\partial t} = \int \frac{d^3 q}{(2\pi)^3} \left(\frac{\dot{R}}{R} q^2 \frac{\partial g}{\partial q} + C_D \right)$$
$$= -4\rho_\nu \frac{\dot{R}}{R} + \int \frac{d^3 q}{(2\pi)^3} C_D. \quad (7.13)$$

Making the now familiar approximations,

$$\int \frac{d^3 q}{(2\pi)^3} C_D = \iiiint (2\pi)^4 \delta(m_L - q - q' - \bar{q}')$$
$$\times\, \delta^{(3)}(\mathbf{q} + \mathbf{q}' + \bar{\mathbf{q}}') W_D(p; q, q', \bar{q}')$$
$$\times\, f_0(p)(1 - e^{\alpha}) 2q\, dQ\, dQ'\, d\bar{Q}'\, dP, \quad (7.14)$$

and

$$f_0(p) = e^{-[\alpha + \beta E_L(p)]}. \quad (7.15)$$

Thus, using (6.21), we can write (7.14) as

$$\int \frac{d^3 q C_D}{(2\pi)^3} = \langle E_l \Gamma \rangle (n - n_{cl}), \quad (7.16)$$

where

$$\langle E_l \Gamma \rangle = \iiint (2\pi)^4 \delta(m_L - q - q' - \bar{q}')$$
$$\times\, \delta^{(3)}(\mathbf{q} + \mathbf{q}' + \bar{\mathbf{q}}') W_D(p; q, q', \bar{q}') q\, dQ\, dQ'\, d\bar{Q}'/2m_L. \quad (7.17)$$

Thus

$$\frac{\partial \rho_\nu}{\partial t} + 4\rho_\nu \frac{\dot{R}}{R} = \langle E_l \Gamma \rangle (n - n_{cl}). \quad (7.18)$$

With the boundary condition†

$$\rho_\nu(t_f) = 0. \quad (7.19)$$

† This is the energy density of the l-particles which are produced in the decay.

Equation (7.18) can be solved to yield,

$$\rho_\nu(t) = \frac{1}{R(t)^4} \int_{t_f}^{t} dt' R(t')^4 \langle E_l \Gamma \rangle [n(t') - n_{cl}(t')]. \tag{7.20}$$

This formula is a more generally correct version of an equation first derived in Dicus, Kolb, and Teplitz (1978) and widely used since. To recover their formula, additional assumptions must be made, which we list below:

1. Suppose that

$$\langle E_l \Gamma \rangle \simeq \langle E_l \rangle \Gamma \simeq \tfrac{1}{3} m_L \Gamma, \tag{7.21}$$

   where Γ is the decay rate, assumed independent of t.
2. In the regime $t_f < t' < t$, $n_{cl}(t') \ll n(t')$. Hence we can use (7.7) in (7.20), dropping the $n_{cl}(t')$ term.
3. Since the decay products of the L are sensibly massless and since they might well dominate ρ after decay and, as each ν carries an energy $\frac{1}{3}m_L$, the universe in this scenario for $t_f < t' < t$ would be expected to be radiation dominated, i.e.,

$$\dot{R}/R = AT^2, \tag{7.22}$$

   or, assuming that

$$RT = \text{constant} \tag{7.23}$$

   during the regime

$$t_f < t < t_p, \tag{7.24}$$

   we have

$$T \sim 1/t^{\frac{1}{2}}. \tag{7.25}$$

Using these various assumptions we can write (7.20) as

$$\rho_\nu = \tfrac{1}{3}\Gamma m_L (t_f/t)^{\frac{3}{2}} n(t_f) \int_{t_f}^{t} dt' (t'/t)^{\frac{1}{2}} e^{-\Gamma(t'-t_f)}. \tag{7.26}$$

There are three such contributions, for both ν and $\bar{\nu}$. Thus we have for the total present energy density due to neutrinos, ρ_p,

$$\begin{aligned}\rho_p &= \rho_{Lp} + \rho_{lp} \\ &= 2m_L n(t_f)(t_f/t_p)^{\frac{3}{2}} \left[e^{-\Gamma(t_p - t_f)} + \Gamma \int_{t_f}^{t_p} dt' (t'/t_p)^{\frac{1}{2}} e^{-\Gamma(t'-t_f)} \right].\end{aligned} \tag{7.27}$$

If one drops the first exponential in the bracket in (7.27) and replaces $n(t_f)$

by $n_{c1}(t_f)$ one has the expression – Eq. (19)–given in Dicus, Kolb, and Teplitz (1978).† We will now discuss these differences.

As Dicus *et al.* point out, the use of $n_{c1}(t_f)$ is an approximation in (7.27). How good an approximation is it likely to be? We can answer this straightforwardly by taking the ratio of $\hat{G}(y_f)$ as given by using (6.119) and $\hat{G}_0(y_f)$ given by (6.103). Thus

$$\frac{\hat{G}_0(y_f)}{\hat{G}(y_f)} = \frac{ce^{-1/y_f}1/y_f^{\frac{3}{2}}}{\{\ln[(\hat{\lambda}/4\xi(3))(\frac{1}{2}\pi)^{\frac{1}{2}}] + \gamma\}(1/\lambda)} \tag{7.28}$$

with $\gamma \simeq 0.6$, and – see (6.110) –

$$c = \frac{1}{2\xi(3)}(\tfrac{1}{2}\pi)^{\frac{1}{2}} \simeq 0.5.$$

From the discussion leading to (6.110) we have

$$y_f \simeq \frac{1}{\ln\hat{\lambda}}. \tag{7.29}$$

Thus (7.28) takes the approximate form,

$$\frac{\hat{G}_0(y_f)}{\hat{G}(y_f)} \simeq \frac{c(\ln\hat{\lambda})^{\frac{3}{2}}}{\ln[(\hat{\lambda}/4\xi(3))(\frac{1}{2}\pi/\ln\hat{\lambda})^{\frac{1}{2}}] + \gamma}. \tag{7.30}$$

If, using the same input as in the last chapter, we take $\ln\hat{\lambda} \simeq 18$, we have

$$\frac{\hat{G}_0(y_f)}{\hat{G}(y_f)} \simeq 2.5. \tag{7.31}$$

This is similar to the error estimate given in Dicus *et al.* But it is clear that there is no need to use the approximate form of $n(t_f)$ in (7.27). It is just as simple to use the essentially correct

$$n(t_f) = \hat{G}(t_f)n_\gamma(t_f), \tag{7.32}$$

where $\hat{G}(t_f)$ is given by (6.119) and $n_\gamma(t_f)$ can be found using (6.98) with the equivalent temperature given by

$$\frac{T_f}{m_L} = \frac{1}{\ln\hat{\lambda}}. \tag{7.33}$$

† In this paper Dicus *et al.* replace the factor $(t_f/t_p)^{\frac{3}{2}}$ by $(T_\nu/T_f)^3$. The presence of T_ν here is a reflection of the fact that in this scenario the energy density, ρ, is dominated by massless neutrinos after decay and these do not share the photon heating.

The second difficulty with (7.27), as written, was also noted by Dicus, Kolb, and Teplitz. If one takes the limit $\Gamma \to 0$ in this equation one returns to our previous example, but with the important difference that the time dependence here is that of a radiation dominant universe. This is not likely to be correct in this limit since the massive L particles will very likely dominate ρ. The generally correct way to write (7.27) is

$$\rho_p = 2m_L n(t_f)[R(t_f)/R(t_p)]^3 \times \left\{ e^{-\Gamma(t_p - t_f)} + \Gamma \int_{t_f}^{t_p} [R(t')/R(t_p)] e^{-\Gamma(t' - t_f)} \, d\bar{c}' \right\}, \tag{7.34}$$

where $R(t)$ is determined by the Einstein equation

$$(\dot{R}/R)^2 = \tfrac{1}{3} 8\pi G_N \rho. \tag{7.35}$$

In the case envisioned by Dicus, Kolb, and Teplitz it is assumed that

$$\Gamma(t_p - t_f) \gg 1, \tag{7.36}$$

which means that for most of its history the universe is radiation dominated and that

$$\rho_p \simeq 2m_L n(t_f)(t_f/t_p)^{\frac{3}{2}} \times \Gamma \int_{t_f}^{t_p} dt'(t'/t_p)^{\frac{1}{2}} e^{-\Gamma(t' - t_f)}. \tag{7.37}$$

They evaluate this integral numerically.† Below we give a partial list of their results. They are obtained by demanding that $\rho_p \leq \rho_c$, where they take as a reasonable value of ρ_c

$$\rho_c = 5 \times 10^{-3} \, \text{MeV/cm}^3 \tag{7.38}$$

to be compared with (6.183). This leads to Table 7.1, with $t = 1/\Gamma$. It is interesting that at l-masses of 7.2×10^3 MeV and 4.7×10^{-5} MeV they recover the fact that the L can be stable without violating the bound set by (7.38). These numbers are in reasonable agreement with our estimates of the last chapter. It is clear from the table that a variety of masses and lifetimes can be accommodated. We wish to turn next to the question of entropy production in this decay.

We can carry the discussion out in the spirit of the work leading to (6.44). We assume that the L particles are represented by, for each spin degree of

† If $t_f \ll 1/\Gamma \ll t_p$ we can drop the positive exponential in (7.37) and extend the limits to $0 \leq t' \leq \infty$ to arrive at the approximate expression

$$\rho_p \simeq \pi^{\frac{1}{2}} m_L n(t_f)(t_f/t_p)^{\frac{3}{2}} [1/(\Gamma \tau_p)^{\frac{1}{2}}].$$

Table 7.1.

Neutrino mass (MeV)	Upper limit of t/t_p
1.00×10^{3}	2.13×10^{-4}
1.00×10^{2}	6.55×10^{-8}
1.00×10^{1}	1.19×10^{-10}
1.00×10^{0}	7.63×10^{-10}
1.00×10^{-1}	7.30×10^{-8}
1.00×10^{-2}	7.30×10^{-6}
1.00×10^{-3}	7.30×10^{-4}
1.00×10^{-4}	7.30×10^{-2}

freedom,

$$f \simeq e^{-(\alpha+\beta E_L)}, \tag{7.39}$$

while the l particles are represented by

$$g \simeq e^{-\beta q}. \tag{7.40}$$

Thus the entropy currents, S^μ, have divergences,

$$\begin{aligned} S^\mu_{L;\mu} \simeq \iiiint &[\alpha + \beta E(\mathrm{p})](2\pi)^4 \\ &\times \delta(m_L - q - q' - \bar{q}')\delta^{(3)}(\mathbf{q} + \mathbf{q}' + \bar{\mathbf{q}}') \\ &\times W(p; q, q', \bar{q}')(1 - e^{-\alpha})e^{-\beta E(p)} 2\,dP\,dQ\,dQ'\,d\bar{Q}', \end{aligned} \tag{7.41}$$

while

$$\begin{aligned} S^\mu_{l;\mu} \simeq -\iiiint &\beta(q + q' + \bar{q}')(2\pi)^4 \delta(m_L - q - q' - \bar{q}') \\ &\times \delta^{(3)}(\mathbf{q} + \mathbf{q}' + \bar{\mathbf{q}}')W(p; q, q', \bar{q}')(1 - e^{-\alpha})e^{-\beta(q+q'+\bar{q}')} \\ &\times 2\,dP\,dQ\,dQ'\,d\bar{Q}'. \end{aligned} \tag{7.42}$$

Thus

$$\begin{aligned} S^\mu_{;\mu} = S^\mu_{L;\mu} + S^\mu_{l;\mu} \simeq \alpha \iiiint &(2\pi)^4 \delta(m_L - q - q' - \bar{q}') \\ &\times \delta^{(3)}(\mathbf{q} + \mathbf{q}' + \bar{\mathbf{q}}')W_0(p; q, q', \bar{q}')(1 - e^{-\alpha})e^{-\beta E(p)} \\ &\times 2\,dP\,dQ\,dQ'\,d\bar{Q}' \\ = \alpha\langle\Gamma\rangle(n_{el} - n) = \frac{\alpha}{R^3}&\frac{\partial}{\partial t}(R^3 n). \end{aligned} \tag{7.43}$$

Since, in this regime, $\partial(R^3 n)/\partial t$ is negative, α must also be negative. If we compare (6.21) and (6.22) with (6.7) we find, ignoring constants and slowly varying terms, that, calling the decay temperature T_d,

$$\alpha \simeq \left(\frac{T_d}{T}\right)^2 - \frac{m_L}{T} \simeq -\frac{m_L}{T}, \tag{7.44}$$

with our assumption that

$$T_d \ll m_L.$$

So α will be negative, and the treatment consistent, so long as

$$T > \frac{T_d}{m_L} T_d.$$

It is clear that α, during this regime, is large. This reflects the fact that the decaying L particles are well out of equilibrium. Nevertheless the net entropy generated may be small, reflecting the fact – see (7.7) – that $\partial(R^3 n)/\partial t$ is a decaying exponential. We may next turn to the question of whether these decaying particles heat the medium.†

Our starting point – see Appendix D – is the Einstein equation

$$\frac{\partial U}{\partial t} = -3R^3 P \frac{\dot{R}}{R}, \tag{7.45}$$

where U is the *total* energy and P the total pressure. In addition to the L and l particles there will be, in any realistic scenario, a background fluid of sensibly massless particles at, we assume, a common temperature among themselves and the L and l particles. We may write

$$U \simeq m_L N_L + \tfrac{3}{2} T N_L + 3(N_l + N_r)T, \tag{7.46}$$

where N_L, N_l, and N_r are the number of L, l, and background particles, respectively. In this regime we suppose that

$$\dot{N}_r = 0, \tag{7.47}$$

and

$$\dot{N}_l + \dot{N}_L = 0. \tag{7.48}$$

Thus

$$\begin{aligned} \dot{U} &= (m_L + \tfrac{3}{2}T)\dot{N}_L + 3T\dot{N}_l + (\tfrac{3}{2}N_L + 3N_l + 3N_r)\dot{T} \\ &= (m_L - \tfrac{3}{2}T)\dot{N}_L + (\tfrac{3}{2}N_L + 3N_l + 3N_r)\dot{T}, \end{aligned} \tag{7.49}$$

† Scherrer and Turner (1985) considered this question and concluded that, under certain assumptions, these decaying particles have no heating effect.

while, using the approximate equation of state,

$$P_L = Tn_L, \tag{7.50}$$

$$R^3P = TN_L + (N_l + N_r)T \tag{7.51}$$

where the only parentheses in (7.51) come from the massless particle relation

$$\tfrac{1}{3}U = R^3P. \tag{7.52}$$

Thus we find that (7.45) produces the relationship

$$\frac{\dot{T}}{T} = \frac{(\dot{R}/R) + \frac{1}{3}\Gamma(N_L/N)(m_L/T)}{1 - \frac{1}{2}N_L/N}, \tag{7.53}$$

where N is the conserved quantity

$$N = N_L + N_l + N_r. \tag{7.54}$$

Since N_L falls exponentially to zero with an e-folding time of $1/\Gamma$ the system will relax rapidly to one characterized by the massless, entropy conserving temperature dependence

$$TR = \text{constant}.$$

However, depending on the relative magnitudes of the terms in the numerator there is a transient phase in which the plasma might have a temperature increase or a decrease with a more complicated dependence.[†]

In our previous work we have assumed that the l particles once created in the L-decays come rapidly into thermal equilibrium and are describable by Boltzmann distributions. It is this assumption we now wish to examine.[‡] The point is that, due to the conservation laws, the momentum distribution at the time of decay – t_d – of the l particles will not be Boltzmann's but will have peaks in momentum space reflecting the conservation of momentum and energy. To make the principles clear we shall study a model system in which the L particle is a boson which decays into two massless bosons – l particles – at time t_d. We call the L-density at this time ρ_0 and, if

[†] In the examples considered by Scherrer and Turner (1985) the temperature always decreases.

[‡] This problem was treated for the momentum distribution of the Boltzmann functions by S. Dodelson, unpublished. We thank S. Dodelson and G. Feinberg for helpful discussions, and S. Dodelson for his unpublished manuscript. The treatment we give here was done in collaboration with E. Weinberg.

the L particles are highly nonrelativistic, then

$$\rho_0 = m_L n_L(t_d). \tag{7.55}$$

The l-number density at this time is given by

$$n_l(t_d) = 2\rho_d/m_L. \tag{7.56}$$

We have assumed that all the L particles have decayed at time t_d into two l particles. We shall assume that, for $t > t_d$, the l particles interact only among themselves by elastic collisions which are l-number conserving so that

$$(R_d)^3 n_l(t_d) = R^3 n_l(t), \tag{7.57}$$

where $t > t_d$. If we call g the Boltzmann function of an l we have for $t > t_d$

$$\begin{aligned}\frac{\partial g}{\partial t} - \frac{\dot{R}}{R} q \frac{\partial g}{\partial q} &= C(E) \\ &= \frac{1}{q} \iiint (2\pi)^4 \delta^{(4)}(q + q' - q_1 - q_2) W(q, q'; q_1, q_2) \\ &\quad \times [g(q_1)g(q_2) - g(q)g(q')] \frac{d^3q' d^3q_1 d^3q_2}{(2(2\pi)^3)^3 q' q_1 q_2}.\end{aligned} \tag{7.58}$$

The equilibrium function, with $\beta \sim R$,

$$g_0 = e^{-(\alpha + \beta q)}, \tag{7.59}$$

is a solution to (7.58) but it does not have the initial condition we are interested in. Instead, we consider a function of the form

$$g(q, R(t)) = N(t) \frac{R_d}{R} \delta\left(q - \frac{R_d}{R} \bar{q}\right) \bar{q}. \tag{7.60}$$

Here, to save writing, we have defined

$$\bar{q} = \tfrac{1}{2} m_L. \tag{7.61}$$

The initial conditions are

$$f(q, R)_{t=t_d} = 0 \tag{7.62}$$

and

$$\int \frac{d^3q}{(2\pi)^3} g(q, R)\bigg|_{t=t_d} = \frac{1}{2\pi^2} \bar{q}^3 N(t_d) = n_d. \tag{7.63}$$

Thus the Boltzmann equation becomes

$$\dot{N}\frac{R_d}{R}\delta\left(q-\frac{R_d}{R}\bar{q}\right)+\frac{\partial f}{\partial t}-\frac{\dot{R}}{R}q\frac{\partial f}{\partial q}=C(E). \tag{7.64}$$

The essential ingredient in $C(E)$ is given by

$$\begin{aligned}[g(q_1)g(q_2)-g(q)g(q')] &= N^2\bar{q}^2\left(\frac{R_d}{R}\right)^2 \\ &\quad\times\left[\delta\left(q_1-\frac{R_d}{R}\bar{q}\right)\delta\left(q_2-\frac{R_d}{R}\bar{q}\right)\right. \\ &\quad\left.-\delta\left(q-\frac{R_d}{R}\bar{q}\right)\delta\left(q'-\frac{R_d}{R}\bar{q}\right)\right]+N\frac{R_d}{R}\bar{q} \\ &\quad\times\left\{\left[\delta\left(q_1-\frac{R_d}{R}\bar{q}\right)f(q_2)+(q_1\to q_2)\right]\right. \\ &\quad\left.-\left[\delta\left(q-\frac{R_d}{R}\bar{q}\right)f(q')+(q\to q')\right]\right\} \\ &\quad+[f(q_1)f(q_2)-f(q)f(q')].\end{aligned} \tag{7.65}$$

What we shall do is to expand (7.64) in powers of the time, t. We shall consider only the first term in this expansion to see how the system begins to depopulate the δ-function distribution. We assume that, as f begins its expansion only in $O(t)$, we can drop all but the first term in (7.65). We are also, during this time interval, going to ignore the expansion of the universe since we shall suppose that $1/t_1$, the l particles collision rate, is much greater than $\dot{R}/R$. Finally, we shall assume that $W(q,q';q_1,q_2)$ is a constant. If we study $g(q)$ for $q\neq\bar{q}$ – recall with our approximations $R_d\simeq R$ – we find that to this order,

$$\begin{aligned}\frac{\partial f}{\partial t}=\frac{1}{(2\pi)^3}\frac{N^2\bar{q}^2W}{8q}\iiint&\delta(q_1-\bar{q})\delta(q_2-\bar{q}) \\ &\times\frac{\delta^{(4)}(q+q'-q_1-q_2)}{q'q_1q_2}\,\mathrm{d}^3q'\,\mathrm{d}^3q_1\,\mathrm{d}^3q_2,\end{aligned} \tag{7.66}$$

where the right-hand side of (7.66) is independent of t. We can evaluate the integral

$$\begin{aligned}I(q)=\iiint\frac{\mathrm{d}^3q'\,\mathrm{d}^3q_1\,\mathrm{d}^3q_2}{q'q_1q_2}&\delta(q_1-\bar{q})\delta(q_2-\bar{q}) \\ &\times\delta^{(4)}(q+q'-q_1-q_2)\,\mathrm{d}^3q'\,\mathrm{d}^3q_1\,\mathrm{d}^3q_2\end{aligned} \tag{7.67}$$

$$= \frac{2(2\pi)^2}{q}(2\bar{q} - q), \quad 2\bar{q} > q > \bar{q}$$

$$= 2(2\pi)^2, \qquad q < \bar{q}.$$

Here, $2\bar{q}$ is the kinematic limit. Thus

$$f(q) = \frac{Wt}{4(2\pi)^3} N^2 \bar{q}^2 \left(\frac{2\bar{q}}{q^2} - \frac{1}{q} \right), \quad 2\bar{q} > q > \bar{q} \tag{7.68}$$

$$= \frac{Wt}{4(2\pi)^3} N^2 \frac{\bar{q}^2}{q}, \qquad q < \bar{q}.$$

This function is continuous at $q = \bar{q}$. As one might expect from the symmetry of the problem, at time t the relative number density of 1 particles with momenta less than $\bar{q}$ – excluding the δ function peak – is the same as the number with $\bar{q} < q < 2\bar{q}$. This number density is given by

$$n_{q \leqslant \bar{q}} = n_{q \leqslant q \leqslant 2\bar{q}} = \frac{1}{128\pi^5} WtN^2 \bar{q}^4. \tag{7.69}$$

The δ-function peak, N, is reduced by a factor of

$$N(t) = N - \frac{WN^2 \bar{q} t}{32\pi^3}. \tag{7.70}$$

This comes from evaluating an integral of the form, where $h(q)$ is arbitrary,

$$I'(q) = \iiint \delta^{(4)}(q + q' - q_1 - q_2) \times h(q') \frac{\mathrm{d}^3 q_1 \, \mathrm{d}^3 q_2 \, \mathrm{d}^3 q'}{q' q_1 q_2} \tag{7.71}$$

$$= \frac{8\pi^2}{q} \int_0^\infty \mathrm{d}q' q' h(q').$$

We next evaluate $n(t)$, where

$$n(t) = \int \frac{\mathrm{d}^3 p}{(2\pi)^3} g(q) = \frac{1}{2\pi^2} \int_0^\infty \mathrm{d}q q^2 \left(N - \frac{WN^2 \bar{q} t}{32\pi^3} \right) \bar{q} \delta(q - \bar{q}) \tag{7.72}$$

$$+ \frac{WtN^2}{64\pi^5} \bar{q}^4 = n(t_\mathrm{d}).$$

This is the expression of the conservation of particle number upon ignoring the cosmic expansion. We may also ask what the average energy per particle is after time t; i.e.,

$$\frac{\langle q \rangle}{n} = \frac{1}{n} \int \frac{\mathrm{d}^3 q q}{(2\pi)^3} g(q) = \bar{q}. \tag{7.73}$$

The fact that $\langle q \rangle / n$ does not change in time can be traced to the fact that

$$T^{\mu 0}{}_{;\mu} = \int C(q) q \frac{\mathrm{d}^3 q}{(2\pi)^3} = 0, \tag{7.74}$$

since the collisions are elastic. If we plot $q^2 g(q)$, which represents the relative proportionality for finding momenta in the interval dq, we find, exclusive of the δ-function at $q = \bar{q}$, a linear rise with q from $0 < q < \bar{q}$ followed by a linear falloff in the interval $\bar{q} < q < 2\bar{q}$. We would expect, as we iterated in t, to discover the coefficient of the δ-function disappearing while the function f approaches an exponential. In principle, at least, it is not difficult to include the expansion of the universe. We begin with the equation

$$\frac{\partial}{\partial t} g(q,t) - \frac{\dot{R}}{R} q \frac{\partial}{\partial q} g(q,t) = C(q,t) = \frac{\dot{R}}{R}\left(R \frac{\partial g}{\partial R} - q \frac{\partial g}{\partial q} \right) \tag{7.75}$$

since t is a function of R. We can introduce the variable

$$x = \frac{R_{\mathrm{d}}}{R} \tag{7.76}$$

so

$$\frac{\dot{x}}{x} = -\frac{\dot{R}}{R}. \tag{7.77}$$

If we define Q by the relation

$$Q = xq \tag{7.78}$$

and note that

$$\frac{\mathrm{d}}{\mathrm{d}x} g(Q,x) = \frac{\partial g}{\partial x} + \frac{\partial g}{\partial Q} q = \frac{1}{x}\left(x \frac{\partial g}{\partial x} + q \frac{\partial g}{\partial q} \right), \tag{7.79}$$

along with the statement that

$$x \frac{\partial}{\partial x} = -R \frac{\partial}{\partial R} \tag{7.80}$$

we arrive at the equation

$$\frac{\mathrm{d}}{\mathrm{d}x} g(Q,x) = \frac{1}{\dot{x}} C(Q,x). \tag{7.81}$$

To implement this equation we need an expression for

$$\frac{\dot{x}}{x} = -\frac{\dot{R}}{R} = -\frac{1}{M_{\text{pl}}}(8\pi\rho/3)^{\frac{1}{2}}, \tag{7.82}$$

where

$$\rho = \int \frac{d^3q}{(2\pi)^3}\, q g(q,t) = \frac{1}{x^4}\int \frac{d^3x}{(2\pi)^3}\, x g(Q,x). \tag{7.83}$$

To proceed in a practical way one would use as a first iterate in (7.83)

$$g_0(Q,x) = Nx\bar{q}\delta(q - x\bar{q}) = N\delta\left(\frac{Q}{x\bar{q}} - 1\right) \tag{7.84}$$

so that ρ_0 is given by

$$\rho_0 = \frac{N}{2\pi^2}(x\bar{q})^4. \tag{7.85}$$

This is the program begun by Dodelson in his manuscript of 1987.

Giving unstable neutrinos a decay mode in which a γ-ray is produced opens up a Pandora's box of new possibilities. It is beyond the scope and purpose of this book to review them. In essence, photons produced prior to recombination can distort the black body spectrum, an effect that can be used to set limits on the neutrino lifetime. On the other hand, if the decay occurs after recombination, these photons may appear as observable ultraviolet, x-ray, or γ-ray backgrounds. This again can be used to set limits. The interested reader can consult the literature cited in the beginning of this chapter. We shall next turn to a study of some of the physics of the recombination regime.

8
The recombination regime

It is a surprising fact that the microwave radiation left over from the Big Bang shows, apart from a secular effect attributable to the motion of our local galaxy, an isotropic black body distribution, to an experimental accuracy of about one part in ten thousand. This is surprising because prior to a temperature of 4000 K $\simeq$ 0.3 eV, and after electron–positron annihilation, the remaining electrons, protons, and photons are expanding as an interacting mixture of relativistic and nonrelativistic particles. We know from our previous work that such a plasma cannot, strictly speaking, be in an equilibrium distribution at a single temperature. We would expect a different temperature, for example, for electrons and photons. The question is, how different? The fact we now observe such a highly accurate black body spectrum implies that this temperature difference must be very small, and to demonstrate this is the aim of the present chapter.†

The electron, positron, proton system has associated with it an energy–momentum tensor $T^{\mu\nu}$ which is covariantly conserved; i.e.,

$$T^{\mu\nu}{}_{;\nu} = 0. \tag{8.1}$$

Since we have an interacting gas this conservation is achieved by cancellation among the $T^{\mu\nu}$ divergences of the different species. We will, in the first instance, focus on $T^{\mu\nu}_{e}$, the electron's energy–momentum tensor. In this regime the electrons are highly nonrelativistic so that we can use the nonrelativistic form of

$$T^{\mu\nu}_{e} = \int \frac{d^3p}{(2\pi)^3} \frac{p^\mu p^\nu}{p^0} g(p), \tag{8.2}$$

where $g(p)$ is the Boltzmann function for the electron. This tensor – see, for example, Eq. (5.14) – obeys a divergence condition of the form

$$T^{0\nu}_{e\ ;\nu} = \frac{1}{m} \int \frac{d^3p}{(2\pi)^3} C(E) \frac{p^2}{2m}, \tag{8.3}$$

† This work was done in collaboration with L. Brown and G. Feinberg.

where $C(E)$ is the collision term. We will return shortly to an analysis of $C(E)$. Since we do not assume that $g(p)$ is an equilibrium distribution we need to define what we mean by the electron temperature, T_e. A natural definition is to take

$$\int \frac{d^3p}{(2\pi)^3} \frac{p^2}{2m} g(p) = \tfrac{3}{2} T_e n_e, \tag{8.4}$$

where n_e is the electron number density. This definition reduces to the equilibrium statement when appropriate. Using it we can write

$$T_e^{00} = n_e m + \tfrac{3}{2} T_e n_e, \tag{8.5}$$

while

$$T_e^{ij} = g^{ij} T_e n_e. \tag{8.6}$$

Since, for our metric,

$$T^{0\nu}{}_{;\nu} = \frac{1}{R^3} \frac{\partial}{\partial t} (R^3 T^{00}) + \delta_{ij} R \dot{R} T^{ij} \tag{8.7}$$

we have, in this case,

$$T_e^{0\nu}{}_{;\nu} = \tfrac{3}{2} n_e \left(\dot{T}_e + 2 \frac{\dot{R}}{R} T_e \right). \tag{8.8}$$

Note that if $T_e^{\mu\nu}$ was separately conserved we would have the equilibrium relation $T_e \sim 1/R^2$.

To deal with the right-hand side of (8.3) we must know what to take for $C(E)$. There are several processes which, at least in principle, contribute to $C(E)$. Among them are electron–photon elastic scattering, electron–electron bremstrahlung, and electron–proton bremstrahlung. It turns out, and we shall give some details later, that the most important process, by far, since $n_\gamma \gg n_e$, is electron–photon scattering and we shall concentrate on it. For this process the collision integral takes the form

$$\begin{aligned} C(E) = \int \frac{d^3k}{(2\pi)^3 2k} \int \frac{d^3k'}{(2\pi)^3 2k'} \int \frac{d^3p'}{(2\pi)^3} \frac{1}{2m} |T|^2 \\ \times (2\pi)^4 \delta^{(4)}(p' + k' - p - k) \{[1 + f(k)] f(k') g(p') \\ - [1 + f(k')] f(k) g(p)\}, \end{aligned} \tag{8.9}$$

where $f(k)$ is the photon distribution. It will turn out that, in this regime, $|T|$, the electron–photon scattering amplitude, is effectively independent of energies and angles and can be factored out of the integrals. This, along with the nonrelativistic kinematics, makes the problem tractable. Since the electron is nonrelativistic

$$|\mathbf{p}| = [2mE(p)]^{\frac{1}{2}} \simeq (2mk)^{\frac{1}{2}} \gg k. \tag{8.10}$$

We shall use this as the basis of a Fokker–Planck-like expansion of the delta-function. Thus

$$\delta^{(3)}(\mathbf{p}' + \mathbf{k}' - \mathbf{p} - \mathbf{k}) \simeq \delta^{(3)}(\mathbf{p}' - \mathbf{p}) + (\mathbf{k}' - \mathbf{k}) \cdot \frac{\partial}{\partial \mathbf{p}'} \delta^{(3)}(\mathbf{p}' - \mathbf{p})$$

$$+ \tfrac{1}{2}(\mathbf{k}' - \mathbf{k}) \cdot \frac{\partial}{\partial \mathbf{p}'} (\mathbf{k}' - \mathbf{k}) \cdot \frac{\partial}{\partial \mathbf{p}'} \delta^{(3)}(\mathbf{p}' - \mathbf{p}) + \cdots. \tag{8.11}$$

The first term in the expansion gives, for the full δ-function,

$$\delta^{(3)}(\mathbf{p}' - \mathbf{p})\delta[E(p') + k' - E(p) - k] = \delta^{(3)}(\mathbf{p}' - \mathbf{p})\delta(k' - k). \tag{8.12}$$

If we put this in (8.9) the term vanishes. The next term in (8.11) effectively vanishes because of the isotropy of the integrand. The last term can be evaluated by using the angular average identity

$$\langle(\mathbf{k}' - \mathbf{k}) \cdot \mathbf{a}(\mathbf{k}' - \mathbf{k}) \cdot \mathbf{b}\rangle = \tfrac{1}{3}\mathbf{a} \cdot \mathbf{b}(k^2 + k'^2). \tag{8.13}$$

Thus

$$\delta^{(4)}(p' + k' - p - k) \simeq \tfrac{1}{6}(k^2 + k'^2) \times \delta\left(\frac{p'^2}{2m^2} + k' - k - \frac{p^2}{2m^2}\right)$$

$$\times \frac{\partial}{\partial \mathbf{p}'} \cdot \frac{\partial}{\partial \mathbf{p}'} \delta^{(3)}(\mathbf{p}' - \mathbf{p}). \tag{8.14}$$

If we imagine inserting a function of p' in the δ-function integration in (8.14) and integrating by parts twice we have effectively

$$\delta^{(4)}(k' + p' - k - p) \simeq \tfrac{1}{6}(k'^2 + k^2)\delta^{(3)}(\mathbf{p}' - \mathbf{p})$$

$$\times \frac{\partial}{\partial \mathbf{p}'} \cdot \left(\frac{\mathbf{p}'}{m} \frac{\partial}{\partial k'} \delta(k' + p' - k - p)\right.$$

$$\left. + \frac{\partial}{\partial \mathbf{p}'} \delta(k' + p' - k - p)\right)$$

$$= \tfrac{1}{6}(k'^2 + k^2)\delta^{(3)}(\mathbf{p}' - \mathbf{p})\left(\frac{3}{m} \frac{\partial}{\partial k'} \delta(k' - k)\right.$$

$$+ \frac{p'^2}{m^2} \frac{\partial^2}{\partial k'^2} \delta(k' - k) + 2 \frac{\partial}{\partial k'} \delta(k' - k) \frac{\mathbf{p}'}{m} \cdot \frac{\partial}{\partial \mathbf{p}'}$$

$$\left. + \frac{\partial}{\partial \mathbf{p}'} \cdot \frac{\partial}{\partial \mathbf{p}'} \delta(k' - k)\right). \tag{8.15}$$

Replacing $g(p)$ by $g(E)$, with

$$E = \frac{p^2}{2m}, \tag{8.16}$$

so that

$$\mathbf{p} \cdot \frac{\partial g}{\partial \mathbf{p}} = 2E \frac{\partial g}{\partial E}, \tag{8.17}$$

and

$$\frac{\partial}{\partial \mathbf{p}} \cdot \frac{\partial g}{\partial \mathbf{p}} = \frac{3}{m} \frac{\partial g}{\partial E} + \frac{2E}{m} \frac{\partial^2 g}{\partial E^2}, \tag{8.18}$$

we have, finally,

$$\begin{aligned}\delta^{(4)}(p' + k' - p - k) \simeq \tfrac{1}{6}(k'^2 + k^2)\delta^{(3)}(\mathbf{p}' - \mathbf{p}) \\ \times \left[\frac{3}{m}\left(\frac{\partial}{\partial k'} \delta(k' - k) + \delta(k' - k)\frac{\partial}{\partial E}\right)\right. \\ + \frac{E}{m}\left(2\frac{\partial^2}{\partial k'^2}\delta(k' - k) + 4\frac{\partial}{\partial k'}\delta(k' - k)\frac{\partial}{\partial E}\right. \\ \left.\left. + 2\delta(k' - k)\frac{\partial^2}{\partial E^2}\right)\right].\end{aligned} \tag{8.19}$$

It is now straightforward, albeit tedious, to substitute this expression into (8.9). We write $C(E)$ in the form

$$C(E) = A_0 g(E) + A_1 \frac{\partial g(E)}{\partial E} + E\left(A_2 g(E) + A_3 \frac{\partial g(E)}{\partial E} + A_4 \frac{\partial^2 g(E)}{\partial E^2}\right), \tag{8.20}$$

where,

$$A_0 = \frac{2}{m^2}|T|^2 \frac{1}{2\pi}\int \frac{\mathrm{d}^3 k}{(2\pi)^3} \frac{k}{2} f(k), \tag{8.21}$$

$$A_1 = \frac{1}{2m^2}|T|^2 \frac{1}{2\pi}\int \frac{\mathrm{d}^3 k}{(2\pi)^3} \frac{k^2}{2} f(k)[1 + f(k)], \tag{8.22}$$

$$A_2 = 0, \tag{8.23}$$

$$A_3 = \tfrac{2}{3}A_0, \tag{8.24}$$

$$A_4 = \tfrac{2}{3}A_1, \tag{8.25}$$

while, with e the electric charge,

$$|T|^2 = \tfrac{4}{3}e^4 = \tfrac{64}{3}\pi^2\alpha^2. \tag{8.26}$$

To proceed, we need to define the photon temperature T_γ. Note that we have the identity

$$\frac{\mathrm{d}}{\mathrm{d}k}\frac{1}{\mathrm{e}^{\beta k}-1} = \frac{-\mathrm{e}^{\beta k}}{(\mathrm{e}^{\beta k}-1)^2}\,\beta = -\beta f_0(k)[1 + f_0(k)], \tag{8.27}$$

where $f_0(k)$ is the equilibrium distribution

$$f_0(k) = \frac{1}{\mathrm{e}^{\beta k}-1}. \tag{8.28}$$

Thus,

$$\int \frac{\mathrm{d}^3k}{(2\pi)^3}\,k^2 f_0(k)[1 + f_0(k)] = \frac{4}{\beta}\int \frac{\mathrm{d}^3k}{(2\pi)^3}\,k f_0(k). \tag{8.29}$$

Hence, in general, we can define T_γ by the relation

$$\int \frac{\mathrm{d}^3k}{(2\pi)^3}\,k^2 f(k)[1 + f(k)] = 4T_\gamma \int \frac{\mathrm{d}^3k}{(2\pi)^3}\,k f(k) = 2T_\gamma\rho_\gamma, \tag{8.30}$$

where ρ_γ is the photon energy density, counting both polarizations. The Thompson cross section, $\sigma_{\text{Thom.}}$, is defined by

$$\sigma_{\text{Thom.}} = \frac{8\pi\alpha^2}{3m^2}. \tag{8.31}$$

Thus

$$A_0 = 2\sigma_{\text{Thom.}}\rho_\gamma \tag{8.32}$$

and

$$A_1 = 2\sigma_{\text{Thom.}}T_\gamma\rho_\gamma, \tag{8.33}$$

which means that $\sigma_{\text{Thom.}}\rho_\gamma$ is a common term in $C(E)$. Thus $C(E)$ can be written as

$$C(E) = 2\sigma_{\text{Thom.}}\rho_\gamma\left[g(E) + T_\gamma\frac{\partial g(E)}{\partial E} + \tfrac{2}{3}E\left(\frac{\partial g(E)}{\partial E} + T_\gamma\frac{\partial^2 g(E)}{\partial E^2}\right)\right]. \tag{8.34}$$

But note that

$$\tfrac{2}{3}\frac{1}{E^{\frac{1}{2}}}\frac{\partial}{\partial E}\left[E^{\frac{3}{2}}\left(g(E) + T_\gamma\frac{\partial g(E)}{\partial E}\right)\right] = \left(g(E) + T_\gamma\frac{\partial g(E)}{\partial E}\right) + \tfrac{2}{3}E\left(\frac{\partial g(E)}{\partial E} + T_\gamma\frac{\partial^2 g(E)}{\partial E^2}\right). \tag{8.35}$$

Thus[†]

$$C(E) = \tfrac{2}{3}\frac{\sigma_{\text{Thom.}}\rho_\gamma}{E^{\frac{1}{2}}}\frac{\partial}{\partial E}\left[E^{\frac{3}{2}}\left(g(E) + T_\gamma \frac{\partial g(E)}{\partial E}\right)\right] \tag{8.36}$$

or, returning to (8.3),

$$\begin{aligned}
T^{0\nu}_{\mathrm{e}\ ;\nu} &= 2\sigma_{\text{Thom.}}\rho_\gamma \frac{1}{3\pi^2}\int_0^\infty \mathrm{d}E E^{\frac{1}{2}}(2mE)^{\frac{1}{2}}E^{\frac{1}{2}} \\
&\quad\times \frac{\partial}{\partial E}\left[E^{\frac{3}{2}}\left(g(E) + T_\gamma \frac{\partial g(E)}{\partial E}\right)\right] \\
&= -\frac{2}{3\pi^2}\sigma_{\text{Thom.}}\rho_\gamma(2m)^{\frac{1}{2}}\int_0^\infty \mathrm{d}E E^{\frac{3}{2}}\left(g(E) + T_\gamma \frac{\partial g(E)}{\partial E}\right) \\
&= -\frac{2}{3\pi^2}\sigma_{\text{Thom.}}\rho_\gamma(2m)^{\frac{1}{2}}\left[\int_0^\infty \mathrm{d}E E^{\frac{3}{2}}g(E) - \tfrac{3}{2}T_\gamma\int_0^\infty \mathrm{d}E E^{\frac{1}{2}}g(E)\right] \\
&= -\tfrac{4}{3}\frac{\sigma_{\text{Thom.}}}{m}\rho_\gamma\left(\int \frac{\mathrm{d}^3p}{(2\pi)^3}\frac{p^2}{2m}g(p) - \tfrac{3}{2}T_\gamma\int\frac{\mathrm{d}^3p}{(2\pi)^3}g(p)\right).
\end{aligned} \tag{8.37}$$

Referring to (8.4) we can write (8.37) as

$$T^{0\nu}_{\mathrm{e}\ ;\nu} = -2\sigma_{\text{Thom.}}\rho_\gamma n_{\mathrm{e}}\left(\frac{T_{\mathrm{e}} - T_\gamma}{m}\right) = \tfrac{3}{2}n_{\mathrm{e}}\left[\dot{T}_{\mathrm{e}} + 2\frac{\dot{R}}{R}T_{\mathrm{e}}\right]. \tag{8.38}$$

Using the same techniques, one can explicitly verify that

$$T^{0\nu}_{\mathrm{e}\ ;\nu} = -T^{0\nu}_{\gamma\ ;\nu}. \tag{8.39}$$

We can write (8.38) in the form[‡]

$$\dot{T}_{\mathrm{e}} + 2\frac{\dot{R}}{R}T_{\mathrm{e}} = -\tfrac{4}{3}\frac{\sigma_{\text{Thom.}}\rho_\gamma}{m}\{T_{\mathrm{e}} - T_\gamma\} \equiv -\frac{1}{t_{\mathrm{e}}}\{T_{\mathrm{e}} - T_\gamma\}, \tag{8.40}$$

where we have defined

$$\frac{1}{t_{\mathrm{e}}} = \tfrac{4}{3}\frac{\sigma_{\text{Thom.}}\rho_\gamma}{m} = \frac{32\pi}{3}\frac{\alpha^2}{m^3}\rho_\gamma. \tag{8.41}$$

Since we do not expect a large deviation for the photon energy distribution from its equilibrium value – $n_\gamma \gg n_{\mathrm{e}}$ – we can take as a first approximation –

[†] The form $\dfrac{1}{E^{\frac{1}{2}}}\dfrac{\partial}{\partial E}h(E)$ is necessary if $C(E)$ is particle conserving.

[‡] This equation can be found in Peebles (1971).

see (6.64) –

$$\rho_\gamma = \frac{\pi^2}{15} T_\gamma^4. \tag{8.42}$$

Thus we can write

$$\frac{1}{t_e} = \frac{32\pi^3}{45}\alpha^2\left(\frac{T_\gamma}{m}\right)^4 m. \tag{8.43}$$

Since

$$m \simeq 0.91 \times 10^{21}/\mathrm{s}, \tag{8.44}$$

we have

$$\frac{1}{t_e} \simeq 1.1 \times 10^{18}\left(\frac{T_\gamma}{m}\right)^4 \mathrm{s}^{-1}. \tag{8.45}$$

This number is to be compared to the expansion rate $\dot{R}/R$, which we assume to be photon dominated in this regime. Thus,

$$\begin{aligned}\frac{\dot{R}}{R} &= \frac{1}{M_{\mathrm{pl}}}\left(\frac{8\pi}{3}\rho_\gamma\right)^{\frac{1}{2}} = \frac{T_\gamma^2}{M_{\mathrm{pl}}}\left(\frac{8\pi^3}{45}\right)^{\frac{1}{2}} \\ &= m\left(\frac{m}{M_{\mathrm{pl}}}\right)\left(\frac{8\pi^3}{45}\right)^{\frac{1}{2}}\left(\frac{T_\gamma}{m}\right)^2 = 0.9 \times \left(\frac{T_\gamma}{m}\right)^2 \mathrm{s}^{-1}.\end{aligned} \tag{8.46}$$

We may now construct a table which will make the numerical situation clear (Table 8.1).

It is evident from this table that, throughout this regime, $1/t_e \gg \dot{R}/R$. This means, looking at (8.40),

$$\left|\frac{T_e - T_\gamma}{T_e}\right| \simeq t_e\dot{R}/R \ll 1. \tag{8.47}$$

This means that the presence of the massive electrons does not, in any practical way, affect the photon temperature. To all intents and purposes the photons follow an equilibrium distribution into the recombination regime. To make this more quantitative we write down the Boltzmann equation for the photon distribution, i.e.,

$$\left[\frac{\partial}{\partial t} - \frac{\dot{R}}{R}k\frac{\partial}{\partial k}\right]f(k,t) = C'(k). \tag{8.48}$$

We take, in the spirit of our previous work, the Thompson scattering

Table 8.1.

T/m	T/k	t s	$\dot{R}/R\ \mathrm{s}^{-1}$	$\frac{1}{t_e}\ \mathrm{s}^{-1}$
1	6×10^{9}	3	0.9	1.1×10^{18}
10^{-2}	6×10^{7}	3×10^{5}	0.9×10^{-4}	1.1×10^{10}
10^{-4}	6×10^{5}	2×10^{9}	0.9×10^{-8}	1.1×10^{2}
10^{-6}	6×10^{3}	2×10^{13}	0.9×10^{-12}	1.1×10^{-6}

contribution to C'; i.e.,

$$C'(k) = \frac{1}{k}\int \frac{d^3k'}{(2\pi)^3 2k'}\int \frac{d^3p'}{(2\pi)^3 2m}\int \frac{d^3p}{2m}|T|^2 \times (2\pi)^4\delta^{(4)}(p' + k' - p - k)\{[1 + f(k)]f(k')g(p') - [1 + f(k')]f(k)g(p)\}. \tag{8.49}$$

Using the same δ-function expansion as above we find†

$$C'(k) \simeq \tfrac{1}{2}\sigma_{\text{Thom.}}\, n_e \frac{1}{mk^2}\frac{\partial}{\partial k}\left[k^4\left(T_e\frac{\partial f(k)}{\partial k} + [1 + f(k)]f(k)\right)\right] \tag{8.50}$$

which explicitly conserves photon number. Since, as we have just shown, in this regime $T_e \simeq T_\gamma$, using (8.27) we see that f_0 makes the left side of (8.48) vanish.‡ This completes the argument that the photons obey a black body distribution prior to recombination.

In the same spirit, it is interesting to ask what the electron distribution is in this regime. Let us make the *anzatz* that it is given by a Maxwell–Boltzmann distribution defined as follows:

$$g = R(t)^{-3}T_e^{-\frac{3}{2}}N_0\exp(-p^2/2mT_e). \tag{8.51}$$

Since we must have

$$R^3\int \frac{d^3p}{(2\pi)^3}g(p) = \text{constant} \tag{8.52}$$

† See also Kamponeets (1957), who did not include the $\dot{R}/R$ term in the Boltzmann equation. With this collision term one can verify that the total $T^{\mu\nu}$ is covariantly conserved.

‡ This assumes that $t_\gamma \sim 1/R$. A chemical potential with $\mu \sim 1/R$ cannot be excluded by this argument.

it follows, upon integration, that N_0 is a constant. We now compute, using (8.40),

$$\left(\frac{\partial}{\partial t} - 2\frac{\dot{R}}{R}E\frac{\partial}{\partial E}\right)g = g\left(\dot{T}_e + 2\frac{\dot{R}}{R}T_e\right)\left(-\frac{3}{2}\frac{1}{T_e} + \frac{E}{T_e^2}\right). \tag{8.53}$$

Referring to (8.53), we can write this as

$$\left(\frac{\partial}{\partial t} - 2\frac{\dot{R}}{R}E\frac{\partial}{\partial E}\right)g = -\tfrac{4}{3}\frac{\sigma_{\text{Thom.}}\rho_\gamma}{m}(T_e - T_\gamma)g \times \left(-\frac{3}{2}\frac{1}{T_e} + \frac{E}{T_e^2}\right). \tag{8.54}$$

But, referring to (8.51),

$$g + T_\gamma \frac{\partial g}{\partial E} = \frac{1}{T_e}(T_e - T_\gamma)g, \tag{8.55}$$

and

$$\frac{\partial g}{\partial E} = -\frac{1}{T_e}g. \tag{8.56}$$

Hence,

$$\left(\frac{\partial}{\partial t} - 2\frac{\dot{R}}{R}E\frac{\partial}{\partial E}\right)g = \frac{2\sigma_{\text{Thom.}}\rho_\gamma}{m}\left[g + T_\gamma\frac{\partial g}{\partial E} + \tfrac{2}{3}E\left(\frac{\partial g}{\partial E} + T_\gamma\frac{\partial^2 g}{\partial E^2}\right)\right], \tag{8.57}$$

which, referring to (8.34) is just the Boltzmann equation to be solved. Thus, in this regime, the electrons are represented by (8.51). It is interesting that in this derivation we have not used the fact that $T_e \simeq T_\gamma$. This means in a situation where these two temperatures differ, but nonrelativistic electron–photon scattering dominates, the electron follows a Maxwell–Boltzmann distribution.

Since, in our regime, electron–photon scattering is sufficiently strong to equilibrate the system, consideration of the other processes is somewhat academic. Nonetheless we may make a few remarks about them. An obvious process is electron–electron Coulomb scattering. This is complicated by the small angle divergence problem which we can treat by using screening in the plasma. If we do this, we find, ignoring constants of $O(1)$, that

$$\frac{1}{t_{\text{Coul.}}} \simeq \left(\frac{n_e}{n_\gamma}\right)^2 \frac{\alpha^3 T_\gamma^3}{m(mT_e)^{\frac{1}{2}}} \simeq 10^{-3}\frac{\dot{R}}{R}. \tag{8.58}$$

Electron–electron bremstrahlung is supressed by the fact that there is no dipole radiation – identical particles. For it, we find that

$$\frac{1}{t_{\mathrm{e-br.}}} \simeq 10^{-14}\frac{T_\gamma^3}{m^2}\left(\frac{T_\gamma}{m}\right)^{\frac{1}{2}} \simeq 10^{-1}\frac{\dot{R}}{R}. \tag{8.59}$$

Finally, there is electron–proton bremstrahlung, for which we find,

$$\frac{1}{t_{\mathrm{p-br.}}} \simeq \frac{1}{t_{\mathrm{e}}}\frac{1}{500}\left(\frac{n_{\mathrm{e}}}{n_\gamma}\right)\left(\frac{m}{T}\right)^{\frac{3}{2}} \simeq 10^{-2}\frac{1}{t_{\mathrm{e}}}, \tag{8.60}$$

where $1/t_{\mathrm{e}}$ is defined by (8.43). We conclude then, that these processes do not make a significant contribution. By the epoch of recombination some of the protons have been incorporated into helium nuclei. Both the protons and the helium nuclei are basically uncoupled to the photons since mass factors suppress the rates. They do not affect the photon temperature. The same considerations explain why the photons emerge from the annihilation regime in a thermal distribution. As Table 8.1 shows, when $T \sim m$, $\dot{R}/R \simeq 10^{-18} \times 1/t_\gamma$ where t_γ is the photon–electron scattering time. This rate is proportional to n_{e} which is dropping by a factor of 10^{10} during annihilation. Even so, the scattering rate remains much greater than $\dot{R}/R$; hence the photon retains its equilibrium distribution.

In this chapter we have only skimmed the surface of the rich physics of the recombination regime.† We would like to conclude the chapter with another example; the so-called "residual ionization" after recombination.‡ During the recombination regime photons are produced via the capture reaction

$$\mathrm{e}^- + \mathrm{p} \rightarrow \mathrm{H} + \gamma, \tag{8.61}$$

where H stands for the hydrogen atom. In fact, capture takes place to an excited state, generally, which then decays to the ground state. This is a nuance we shall ignore in our approximate numerical estimate. In any event, not all the electrons are captured. Let us define the residual number remaining, x, by the equation

$$x = \frac{n_{\mathrm{e}}}{n_{\mathrm{B}}} = \frac{n_{\mathrm{e}}}{n_\gamma}\frac{n_\gamma}{n_{\mathrm{B}}}, \tag{8.62}$$

† For a survey see, for example, Zel'dovich and Sunyaev (1969).

‡ This problem was studied by Zel'dovich and Sunyaev (1969). The work presented here was done in collaboration with L. Brown, with helpful conversations with M. Turner.

where n_B, n_e, n_γ are the baryon, electron, and photon number densities. The quantity x is a function of the temperature, but we will be interested in x_0, the present value of x. We then have another example of a problem to which the methods of Chapter 6, suitably modified, are applicable. Indeed, one of the interests in working out this problem is to see how these modifications are to be made. If we call the quantity n_e simply n, the rate equation is, as usual,

$$\frac{1}{R^3}\frac{\partial}{\partial t}(R^3 n) = (n_{cl}^2 - n^2)\langle \sigma v\rangle, \tag{8.63}$$

or

$$\frac{\partial}{\partial t}\left(\frac{n}{n_\gamma}\right) = \langle \sigma v\rangle T^3\frac{2\xi(3)}{\pi^2}\left(\frac{n_{cl}^2}{n_\gamma^2} - \frac{n^2}{n_\gamma^2}\right), \tag{8.64}$$

where $\langle \sigma v\rangle$ must be specified. In deriving (8.64) we have used the fact that

$$RT = \text{constant}. \tag{8.65}$$

Apart from the new physics of $\langle \sigma v\rangle$, we must also remember that in this regime, at least in the conventional scenario, ρ is *matter* dominated. Thus

$$\rho = m n_B, \tag{8.66}$$

where m is the mass of the proton. Since

$$R^3 n_B = \text{constant}, \tag{8.67}$$

we have the scaling relation

$$\frac{\rho}{\rho_0} = \frac{R_0^3}{R^3} = \frac{T^3}{T_0^3}. \tag{8.68}$$

Thus we can write

$$\frac{\partial}{\partial t} = -\left[\frac{8\pi}{3}\frac{1}{M_{pl}^2}\frac{2\xi(3)}{\pi^2}\left(\frac{\rho}{n_\gamma}\right)_0\right]^{\frac{1}{2}} T^{\frac{5}{2}}\frac{\partial}{\partial T}. \tag{8.69}$$

It is convenient to introduce the variable

$$y = T/\epsilon_H. \tag{8.70}$$

where $\epsilon_H = 13.6$ eV, the ground state binding energy. Hence we may write the transformed version of (8.64) as

$$\frac{\partial}{\partial y}\left(\frac{n}{n_\gamma}\right) = \frac{\langle \sigma v\rangle \epsilon_H^2 y^{\frac{1}{2}} 2\xi(3)/\pi^2}{\left[\frac{16}{3\pi}\xi(3)\frac{\epsilon_H}{M_{Pl}^2}\left(\frac{\rho}{n_\gamma}\right)_0\right]^{\frac{1}{2}}} \times \left[\left(\frac{n}{n_\gamma}\right)^2 - \left(\frac{n_{Cl}}{n_\gamma}\right)^2\right]. \tag{8.71}$$

Hence the quantity $\hat{\lambda}$, introduced in Chapter 6 becomes, for this problem,

$$\hat{\lambda} = \frac{\langle \sigma v \rangle \epsilon_{\mathrm{H}}^2 y^{\frac{1}{2}} 2\xi(3)/\pi^2}{\left[\dfrac{16}{3\pi}\xi(3)\dfrac{\epsilon_{\mathrm{H}}}{M_{\mathrm{Pl}}^2}\left(\dfrac{\rho}{n_\gamma}\right)_0\right]^{\frac{1}{2}}}. \tag{8.72}$$

For σ we use the approximate expression† for capture into the ground state

$$\sigma = \frac{1.96\pi^2\alpha}{m_{\mathrm{e}}^2}\frac{\epsilon_{\mathrm{H}}^2}{E(E-\epsilon_{\mathrm{H}})}. \tag{8.73}$$

Here,

$$E = \tfrac{1}{2}m_{\mathrm{e}}v^2 + \epsilon_{\mathrm{H}}. \tag{8.74}$$

We shall set

$$\tfrac{1}{2}m_{\mathrm{e}}v^2 = \tfrac{3}{2}T. \tag{8.75}$$

We shall apply this expression in a regime in which $T/\epsilon_{\mathrm{H}} < 1$. Hence we can write $\langle \sigma v \rangle$ in the form, using (8.75) to eliminate v,

$$\langle \sigma v \rangle \simeq \frac{4\pi^2\alpha}{3^{\frac{1}{2}}}\frac{1}{m_{\mathrm{e}}^2}\left(\frac{\epsilon_{\mathrm{H}}}{m_{\mathrm{e}}}\right)^{\frac{1}{2}}\frac{1}{y^{\frac{1}{2}}}. \tag{8.76}$$

We see that, with these approximations, the y dependence in $\hat{\lambda}$ disappears and we can use the results of Chapter 6 directly, i.e., from (6.176)

$$\left(\frac{n}{n_\gamma}\right)_0 \simeq \frac{1}{\hat{\lambda}}\ln\left(\frac{\hat{\lambda}}{4(\ln\hat{\lambda})^{\frac{1}{2}}}\right). \tag{8.77}$$

To determine $\hat{\lambda}$ we shall take, as an example,

$$\left(\frac{n_{\mathrm{B}}}{n_\gamma}\right)_0 \simeq 10^{-10}. \tag{8.78a}$$

and

$$\rho_0 = 5 \times m(n_{\mathrm{B}})_0. \tag{8.78b}$$

With this choice

$$\hat{\lambda} \simeq 5.3 \times 10^{13} \tag{8.79}$$

† See Bethe and Salpeter (1957).

and

$$\left(\frac{n}{n_\gamma}\right)_0 \simeq 5.4 \times 10^{-13} \tag{8.80}$$

while

$$x_0 \simeq 5.4 \times 10^{-3}, \tag{8.81}$$

a surprisingly large result, in general agreement with calculations done by other methods.† We turn next to the matter of helium production.

† M. Turner, private communication.

9
Cosmological helium production

One of the crucial observations that lends credence to the standard Big Bang cosmology is that the universe consists, by weight, of about 25 percent helium. While helium is manufactured in stars, as part of the thermonuclear burning process, it is not in a quantity sufficient to account for this observation. This has led to the speculation that most of this helium has an early universe cosmological origin. The aim of this chapter is to present a simplified model of this process which reproduces the essential elements of the highly sophisticated computer calculations† to within a few percent.

Roughly speaking, the formation of primordial helium occurs as follows. At temperature $T \sim 100$ MeV the energy density of the universe is dominated by sensibly massless particles: electrons, positrons, photons, and neutrinos. There are, in number – as compared to the massless particles – some 10^{-10} neutrons and protons. At these temperatures all of these particles are kept in equilibrium by a combination of weak and electromagnetic interactions. Chemical equilibrium of the baryons and leptons is maintained by the inelastic weak processes

$$\nu_e + \mathrm{n} \leftrightarrow \mathrm{p} + \mathrm{e}^-,$$

$$\mathrm{e}^+ + \mathrm{n} \leftrightarrow \mathrm{p} + \bar{\nu}_e,$$

and

$$\mathrm{n} \leftrightarrow \mathrm{p} + \mathrm{e}^- + \bar{\nu}_e.$$

In the usual treatment – which we shall follow here – it is assumed that all these particles have negligible chemical potentials (see Chapter 6). If the neutrino should have a chemical potential it would affect the results materially. We shall discuss this at the end of the chapter. Therefore, in this equilibrium regime

$$n_n(T)/n_p(T) = \exp(-\Delta m/T), \tag{9.1}$$

† The model was developed in collaboration with L. Brown and G. Feinberg (Bernstein, Brown, and Feinberg, 1988). The full theory can be found, for example, in Peebles (1966).

where Δm is the neutron–proton mass difference,

$$\Delta m = m_n - m_p \simeq 1.29 \text{ MeV}. \tag{9.2}$$

It is convenient to introduce the quantity $X(T)$ defined as

$$X(T) = \frac{n_n(T)}{n_n(T) + n_p(T)}. \tag{9.3}$$

So long as chemical equilibrium is maintained

$$X(T) = X_{eq.}(T) = \frac{1}{1 + \exp(\Delta m/T)}. \tag{9.4}$$

As we shall argue, cosmological helium production occurs when the age of the universe is about four minutes. Since this is a time that is short compared to the *mean* life of the neutron $t \simeq 896 \pm 16$ s, it is a good first approximation to neglect the neutron decay process and its inverse. We shall later in the chapter include this effect. As the universe expands and cools, the chemical equilibrium is broken and, neglecting neutron decay, the quantity $X(T)$ approaches a constant nonzero value as $T \to 0$.

To see how this happens in detail, we study the Einstein equation for which we need to know ρ in this regime. Referring to Eq. 6.64 and Table 6.1 we see that, assuming all the neutrinos are sensibly massless,

$$\rho = \frac{\pi^2}{30} T^4 N_{DF}, \tag{9.5}$$

where, prior to e^+–e^- annihilation,

$$N_{DF} = 10.75. \tag{9.6}$$

At a time t_f, or a temperature T_f, $\dot{R}/R$ becomes comparable to the interaction rates and the baryons become, essentially, decoupled from the leptons. The neutron to proton ratio, and hence X, is frozen at the value

$$X(T_f) \simeq X_{eq}(T_f) \simeq X(0). \tag{9.7}$$

The weak rate, Λ, is given, approximately, by

$$\Lambda \simeq n_{\nu_e} \langle \sigma v \rangle, \tag{9.8}$$

where n_{ν_e} is the density of electron neutrinos. As we argued in Chapter 6, equating Λ and $\dot{R}/R$ produces $T_f \sim 1$ MeV. If we allow N_{DF} to vary, to take into account the possibility of additional neutrino flavors, we have, more precisely,

$$T_f \simeq N_{DF}^{\frac{1}{6}} \times 1 \text{ MeV}. \tag{9.9}$$

In any event, for $T \gtrsim T_f$ the light nuclei D, T, ^{3}He and ^{4}He are kept in thermal and chemical equilibrium by reactions such as

$$n + p \leftrightarrow D + \gamma,$$

$$D + D \leftrightarrow T + p,$$

and

$$T + D \leftrightarrow {}^4H_e + n.$$

The population of these nuclei relative to the leptons and photons is very small. However, as we shall see, once the temperature has fallen below about a thirtieth of the deuteron binding energy, $\epsilon_D = 2.23$ MeV, a temperature which is much smaller than T_f, the nuclear reactions shown above proceed almost entirely to the right and because of the large binding energy of ^{4}He ($\epsilon_{He} \simeq 28.3$ MeV) nearly all of the original neutrons which exist at the temperature T_f are captured in ^{4}He. Thus at the end of the Big Bang nucleosynthesis the ratio of the number of ^{4}He nuclei to the total number of baryons is given by

$$X = \tfrac{1}{2}X(T \simeq 0) \tag{9.10}$$

or, equivalently, the ^{4}He mass fraction, neglecting the tiny effect due to the binding energy, is given by

$$Y = 2X(T \simeq 0). \tag{9.11}$$

Thus, what is needed to determine Y is a theory of $X(T \simeq 0)$, to which we return shortly. Note that the helium mass fraction is related to the exponential of $\Delta m/T_f \sim \Delta m(G_f^2 M_{pl})^{\frac{1}{3}}$. If this pure number were much larger than unity, there would have been no primordial helium production. It is remarkable that this number *is* of order unity since different components of it involve such very different physics: Δm is a nuclear binding energy that is ultimately related to the strong interaction coupling constant, or to the mass scale of strong interactions, while G_f is the strength or mass scale of the weak interaction, while M_{pl} is the mass scale appropriate to the gravitational interaction.

To describe our simplified model of neutron abundance we denote by $\lambda_{pn}(t)$ the rate for the weak processes that convert protons into neutrons, and by $\lambda_{np}(t)$ the rate for the reverse processes. If we write the rate equation in terms of the density ratio, $X(t)$, the factors of R^3 cancel out so there is no $\dot{R}/R$ term in the rate equation. Thus,

$$\frac{dX(t)}{dt} = \lambda_{pn}(t)[1 - X(t)] - \lambda_{np}(t)X(t). \tag{9.12}$$

This rate equation has the solution

$$X(t) = \int_{t_0}^{t} dt' I(t,t')\lambda_{pn}(t') + I(t,t_0)X(t_0), \tag{9.13}$$

where

$$I(t,t') = \exp\left[-\int_{t'}^{t} dt''\,\Lambda(t'')\right], \tag{9.14}$$

and

$$\Lambda(t) = \lambda_{pn}(t) + \lambda_{np}(t). \tag{9.15}$$

The rates λ_{pn} and λ_{np} are very large at early times t when the temperature T is about 100 MeV or so. If the t_0 in (9.13) is such an early time, while t is somewhat later, $I(t,t_0)$ will be very small. Therefore the initial value $X(t_0)$ (which must lie between zero and one) plays no role in (9.13), and can be omitted. For times t somewhat later than t_0, the fast reaction processes wash out the initial condition. For the same reason, the integration in (9.13) is insensitive to the value of t_0. We may simplify the expression by setting $t_0 = 0$, as the integral from $t' = 0$ to $t' = t_0$ is negligible. Hence to a good approximation we may write

$$X(t) = \int_{0}^{t} dt'\, I(t,t')\lambda_{pn}(t'). \tag{9.16}$$

As we will now show, the neutron population is in equilibrium until fairly late times. To see this, and to see how equilibrium is broken, we note that

$$I(t,t') = \frac{1}{\Lambda(t')}\frac{d}{dt'}I(t,t'), \tag{9.17}$$

so that we may integrate by parts to obtain

$$X(t) = \frac{\lambda_{pn}(t)}{\Lambda(t)} - \int_{0}^{t} dt'\, I(t,t')\frac{d}{dt'}\left(\frac{\lambda_{pn}(t')}{\Lambda(t')}\right). \tag{9.18}$$

In the regime, where the total reaction rate, $\Lambda(t)$, is large in comparison to the rate of time variation of the rates, the last term in (9.18) yields a small correction. This correction is exhibited by again performing an integration by parts to give,

$$X(t) \simeq \frac{\lambda_{pn}(t)}{\Lambda(t)} - \frac{1}{\Lambda(t)}\frac{d}{dt}\left(\frac{\lambda_{pn}(t)}{\Lambda(t)}\right). \tag{9.19}$$

With the leptons kept in tight thermal equilibrium by scattering processes, the principle of detailed balance requires that

$$\lambda_{pn}(t) = \exp\left(-\frac{\Delta m}{T(t)}\right)\lambda_{np}(t). \tag{9.20}$$

Thus

$$\frac{\lambda_{pn}(t)}{\Lambda(t)} = \frac{1}{1+\exp(\Delta m/T)} = X_{eq}(T), \tag{9.21}$$

and thus

$$X(t) \simeq \left(1 - \frac{1}{\Lambda(t)}\dot{T}\frac{d}{dT}\right)X_{eq}[T(t)]. \tag{9.22}$$

In this regime we have

$$RT = \text{constant}. \tag{9.23}$$

Hence we have

$$X(t) \simeq X_{eq}[T_{eff}(t)], \tag{9.24}$$

where

$$T_{eff}(t) = \left(1 + \frac{1}{\Lambda}\frac{\dot{R}}{R}\right)T. \tag{9.25}$$

We see, therefore, that $X(t)$ follows its equilibrium value until a time t at which $\dot{R}/R \simeq \Lambda$, the condition one would expect. When $\dot{R}/R$ is of $O(\Lambda)$, $X(t)$ becomes an equilibrium ratio $X_{eq}(T_{eff})$ with an effective temperature $T_{eff}(t)$ that is higher than the temperature of the universe, with $X_{eq}(T_{eff})$ larger than $X_{eq}(T)$. This is the onset of the freezing out of the neutrons.

To go further, we need an explicit form for the rate $\lambda_{np}(t)$. This rate is the sum of the rates of the various inelastic processes which convert n to p, i.e.,

$$\lambda_{np} = \lambda(\nu + n \rightarrow p + e^-) + \lambda(e^+ + n \rightarrow p + \bar{\nu}) + \lambda(n \rightarrow p + \bar{\nu} + e^-). \tag{9.26}$$

We take the formulae for these rates from Weinberg (1972). Thus

$$\lambda(\nu + n \rightarrow p + e^-) = A\int_0^\infty dp_\nu p_\nu^2 p_e E_e(1 - f_e)f_\nu, \tag{9.27}$$

$$\lambda(e^+ + n \rightarrow p + \bar{\nu}) = A\int_0^\infty dp_e p_e^2 p_\nu E_\nu(1 - f_\nu)f_e, \tag{9.28}$$

and

$$\lambda(\mathrm{n} \to \mathrm{p} + \bar{\nu} + \mathrm{e}^-) = A \int_0^{p_0} \mathrm{d}p_e p_e{}^2 p_\nu E_\nu (1 - f_\nu)(1 - f_e). \tag{9.29}$$

Here, A is an effective coupling constant which we shall later eliminate in favor of the free neutron decay rate; so its value need not concern us here. We have called the magnitudes of the neutrino and electron momenta p_ν and p_e with the corresponding energies $E_\nu = p_\nu$ and $E_e = (p_e{}^2 + m_e{}^2)^{\frac{1}{2}}$. In the low energy (compared to the nucleon rest masses) regime which concerns us we can neglect the kinetic energy of the nucleons. Thus p_e in (9.27) is determined by the energy conservation condition

$$E_e = E_\nu + \Delta m, \tag{9.30}$$

with Δm given by (9.2), while E_e in (9.28) is determined by

$$E_\nu = E_e + \Delta m. \tag{9.31}$$

The neutrino energy in (9.29) is determined by

$$E_\nu = \Delta m - E_e \geqslant 0. \tag{9.32}$$

This gives for p_0 in (9.29) the value $p_0 = [(\Delta m)^2 - m_e{}^2]^{\frac{1}{2}}$. The $p^2\,\mathrm{d}p$ factors are phase space factors, while the remaining $p_e{}^2\,\mathrm{d}p_e$ and $p_\nu E_\nu$ factors arise from the transition matrix element. As we argued in Chapter 6, it is a good approximation to take the equilibrium density functions for the leptons in this regime. Thus

$$f_\nu = \frac{1}{\exp(E_\nu/T_\nu) + 1}, \tag{9.33}$$

while

$$f_e = \frac{1}{\exp(E_e/T_e) + 1}. \tag{9.34}$$

During this regime $T_\nu \neq T_e$. The reason is that, at the end of the freezing period for the neutrons, the temperature – nearly identical for ν and e – drops somewhat below $2m_e$, and the electrons and positrons annihilate, heating the photons, but not the neutrinos which are now decoupled. Since the reaction $\mathrm{e}^+ + \mathrm{e}^- \to 2\gamma$ occurs rapidly in comparison with $\dot{R}/R$, the electrons maintain thermal equilibrium with the photons so that $T_e = T_\gamma$, and the overall entropy of the system is conserved. Using the conservation of entropy one can show (see, for example, Weinberg, 1972) that T_ν and T_γ differ by at most 10 percent prior to freeze out. Our first approximation is then to set $T_\nu = T_e = T_\gamma = T$. Within this approximation the rates for the reverse reactions such as $e^- + p \rightarrow n + \nu$ obey the principle of detailed balance so that we have, for example,

$$\lambda(e^- + p \rightarrow n + \nu) = \exp(-\Delta m/T)\lambda(\nu + n \rightarrow p + e^-). \tag{9.35}$$

Adding up all the processes yields the detailed balance statement of (9.19).

The next approximation we make stems from that fact that in this regime T is small compared to Δm and E. Hence we replace the Fermi–Dirac distributions by their Boltzmann equivalents, i.e.,

$$f_{\nu,e} \simeq \exp(-E_{e,\nu}/T). \tag{9.36}$$

We also set

$$1 - f_{\nu,e} \simeq 1, \tag{9.37}$$

since the Boltzmann functions are small in this regime. Hence we can now write

$$\lambda(\nu + n \rightarrow p + e^-) = A\int_0^\infty dp_\nu p_\nu^2 p_e E_e \exp(-E_\nu T), \tag{9.38}$$

$$\lambda(e^+ + n \rightarrow p + \bar{\nu}) = A\int_0^\infty dp_e p_e^2 p_\nu E_\nu \exp(-E_e/T), \tag{9.39}$$

and

$$\lambda(n \rightarrow p + \bar{\nu} + e^-) = A\int_0^{p_0} dp_e p_e^2 p_\nu E_\nu. \tag{9.40}$$

Our final approximation is connected to the last one; in equations (9.38) and (9.39) we neglect the electron mass. In this approximation the first two rates become identical. Inserting $p_e = E_e = \Delta m + E_\nu$ and $p_\nu = E_\nu$ in (9.38) we obtain

$$\begin{aligned}\lambda(\nu + n \rightarrow p + e^-) &= AT^3(4!\,T^2 + 2\cdot 3!\,T\Delta m + 2!\,\Delta m^2) \\ &= \lambda(e^+ + n \rightarrow p + \bar{\nu}).\end{aligned} \tag{9.41}$$

Equation (9.40) is just the decay rate $1/t_n$ for a free neutron. Setting $E_\nu = \Delta m - E_e$ and recalling that $p_0 = [(\Delta m)^2 - m_e{}^2]^{\frac{1}{2}}$ several elementary integrations yield

$$\begin{aligned}1/t_n &= \lambda(n \rightarrow p + \bar{\nu} + e^-) \\ &= \tfrac{1}{5}A[(\Delta m)^2 - m_e{}^2]^{\frac{1}{2}}[\tfrac{1}{6}(\Delta m)^4 - \tfrac{3}{4}(\Delta m)^2 m_e{}^2 - \tfrac{2}{3}m_e{}^4] \\ &\quad + \tfrac{1}{4}Am_e^4\,\Delta m\cosh^{-1}(\Delta m/m_e).\end{aligned} \tag{9.42}$$

Using the numerical values $\Delta m = 1.29$ MeV and $m_e = 0.511$ MeV we obtain

$$1/t_n = 0.0157\ A(\Delta m)^5. \tag{9.43}$$

We shall use this result to provide the scale for the rates in terms of the directly observable mean life of the neutron. Thus we write

$$4A = a(\Delta m)^5/t_n, \tag{9.44}$$

in which a is the pure number

$$a = 255. \tag{9.45}$$

If we neglect, for the moment, the neutron decay rate in $\lambda_{np}(t)$ and introduce the dimensionless temperature variable

$$y = \Delta m/T, \tag{9.46}$$

we can express this part of the rate, which we shall continue to call λ_{np}, as

$$\lambda_{np}(t) = (a/t_n y^5)(12 + 6y + y^2). \tag{9.47}$$

We may note that for $T \gtrsim 1$ MeV, or $y \lesssim 1$, this rate is three orders of magnitude larger than the free neutron decay rate.

We can now compute the neutron abundance, changing variables from the time t to the scaled inverse temperature y. Thus

$$X(y) = \frac{\lambda_{pn}(y)}{\Lambda(y)} - \int_0^y dy' I(y, y') \frac{d}{dy'} \left(\frac{\lambda_{pn}(y')}{\lambda(y')} \right). \tag{9.48}$$

The detailed balance relation gives

$$\lambda_{pn}(y) = e^{-y} \lambda_{np}(y), \tag{9.49}$$

so that

$$\Lambda(y) = (1 + e^{-y}) \lambda_{np}(y). \tag{9.50}$$

The integrating factor now becomes

$$I(y, y') = \exp\left[-\int_{y'}^{y} dy'' \frac{dt''}{dy''} \Lambda(y'') \right]. \tag{9.51}$$

To compute the Jacobian dt''/dy'' we recall that in our regime

$$RT = \text{constant}.$$

This, along with the expression for ρ, the energy density, gives

$$\frac{dT}{dt} = -\left(\frac{4}{45} \frac{\pi^3}{M^2_{pl}} N_{DF} \right)^{\frac{1}{2}} T^3, \tag{9.52}$$

with

$$N_{DF} = 10.75.$$

Hence, we now find that the integrating factor takes the form

$$I(y, y') = \exp[-K(y) + K(y')], \tag{9.53}$$

where

$$K(y) = b \int^{y} dy' \left(\frac{12}{y'^4} + \frac{6}{y'^3} + \frac{1}{y'^2} \right)(1 + e^{-y'}), \tag{9.54}$$

where b is the pure number

$$b = a \left[\frac{45}{4\pi^3 N_{DF}} \right]^{\frac{1}{2}} \frac{M_{pl}}{t_n (\Delta m)^2}. \tag{9.55}$$

It is quite remarkable that the numerical coefficients inside the integration are such as to give a simple closed form

$$K(y) = -b \left[\left(\frac{4}{y^3} + \frac{3}{y^2} + \frac{1}{y} \right) + \left(\frac{4}{y^3} + \frac{1}{y^2} \right) e^{-y} \right]. \tag{9.56}$$

Introducing

$$X_{eq}(y) = \frac{\lambda_{pn}(y)}{\lambda(y)} = \frac{1}{1 + e^{y}}, \tag{9.57}$$

the neutron abundance ratio now reads

$$X(y) = X_{eq}(y) + \int_0^y dy' e^{y'} X_{eq}(y')^2 \exp[K(y') - K(y)]. \tag{9.58}$$

It is a simple matter to numerically compute the integral that appears here for a range of y values. Using the numerical values of the various parameters the dimensionless constant is given by

$$b = \frac{0.823}{(N_{DF})^{\frac{1}{2}}} = 0.251. \tag{9.59}$$

Asymptotically we find that

$$X(y = \infty) = 0.151. \tag{9.60}$$

In what has gone before we have ignored neutron decay. If we include decay, the neutron abundance now becomes

$$\bar{X}(t) = e^{-t/t_n} X[y(t)]. \tag{9.61}$$

To apply (9.61) we must know the time at which the deuterons collide and form into ^{4}He. To determine this time, we note that the neutrons, protons, and deuterons behave as a free nonrelativistic gas with number densities given by

$$n_a = g_a e^{-(\mu_a + m_a)/T} \int \frac{d^3p}{(2\pi)^3} e^{-p^2/2m_a T}$$
$$= g_a e^{-(\mu_a + m_a)/T} \left(\frac{m_a T}{2\pi}\right)^{\frac{3}{2}}. \tag{9.62}$$

Here, g_a is the statistical spin factor given by $g_n = g_p = 2$ and $g_d = 3$, while μ_a and m_a are the chemical potentials and masses, respectively. Since these gases are in chemical equilibrium

$$\mu_D = \mu_n + \mu_p. \tag{9.63}$$

Thus

$$\frac{n_n n_p}{n_D} = \frac{g_p g_n}{g_D} \left(\frac{m_p m_n}{m_D}\right)^{\frac{3}{2}} \left(\frac{T}{2\pi}\right)^{\frac{3}{2}} e^{-\epsilon_b/T}, \tag{9.64}$$

where

$$\epsilon_b = m_p + m_n - m_D, \tag{9.65}$$

is the deuteron binding energy. It is convenient to introduce the photon number

$$n_\gamma = \frac{2\xi(3)}{\pi^2} T^3, \tag{9.66}$$

and the baryon–photon ratio

$$\eta = n_B/n_\gamma. \tag{9.67}$$

Thus (9.64) becomes

$$G_{np} \equiv \frac{n_p n_n}{n_B n_D} = \frac{12\xi(3)}{\pi^{\frac{1}{2}}} \eta \left(\frac{T}{m_p}\right)^{\frac{3}{2}} e^{\epsilon_b/T}. \tag{9.68}$$

Deuterons are formed when

$$G_{np} \simeq 1. \tag{9.69}$$

Since the baryon–photon ratio is very small; i.e.,

$$\eta \simeq 5 \times 10^{-10}, \tag{9.70}$$

this requires that the rapidly varying exponential factor $\exp(\epsilon_{b/T})$ be very large. Solving, we find that the T corresponding to (9.69) is given by

$$T \simeq \frac{\epsilon_b}{33.6} \simeq 0.066\,\text{MeV}. \tag{9.71}$$

To translate this temperature into the corresponding age of the universe is somewhat complicated. (For a detailed discussion see, for example, Bernstein, Brown, and Feinberg, 1988, or Weinberg, 1972.) We wish, in the spirit of the rest of the chapter, to elucidate the essential physics. The connection is, in general, obtained by integrating the Einstein equation (9.52). To do this one must, evidently, know ρ as a function of T. Down to a temperature the order of the electron's rest mass ρ is given by the contributions of sensibly masslesss particles and one need only integrate (9.52) with a constant N of 43/4. Thus to $T \sim m_e$ we find

$$t = \left[\frac{45}{63\pi^3 N_{\text{DF}}}\right]^{\frac{1}{2}} \frac{M_{\text{pt}}}{T^2} = 0.72/(T/\text{MeV})^2\ \text{s}, \tag{9.72}$$

which should be valid until $t \simeq 3$ seconds. Thereafter, the electrons and positrons begin to annihilate. One can argue that the rate for this process is larger than the expansion rate of the universe during this regime so that the electrons and positrons are well-described by their equilibrium distributions which are approximately $\exp(-m/T) \times \exp(-p^2/2mT)$. When these are put into the expression for ρ the electron's contribution to the energy density is multiplied by a factor of $\exp(-m/T)$ and is thus rapidly decreasing throughout this regime. Thus we can, to elucidate the physics, suppose that the electron–positron annihilation occurs instantaneously once the temperature has dropped below the rest mass. Once the electrons are gone, ρ is once again dominated by the massless photons and neutrinos. However, it is not correct to take the electron and photon temperatures to be the same during this regime. By the time of nucleosynthesis we have $T/T_\nu = 1.4$. To take this into account accurately one must (see again Weinberg, 1972) use entropy conservation and do the relevant numerical integral. But, these calculations show that it is a very good approximation in this regime to take $T/T_\nu = 1.4$ throughout. The expanding gas behaves as an independent gas of photons at temperature T and an independent gas of neutrinos at temperature T_ν. Hence the effective number of degrees of freedom in this regime, N_{eff}, is given by the formula

$$N_{\text{eff}} = 2 + (\tfrac{4}{11})^{\frac{4}{3}}\tfrac{7}{4}N_\nu, \tag{9.73}$$

where N_ν is the number of neutrino flavors. With $N_\nu = 3$, $N_{\text{eff}} = 3.36$. It is this value we shall use in evaluating Y. If we put this N_{eff} into (9.72) we find a freezing time $t_c = 300\,\text{s}$. Since Y is given by the equation

$$Y = 2\exp(-t_c/t_n)X(T_F) \tag{9.74}$$

we find, using (9.60), and $t_c = 300\,\text{s}$ along with $\tau = 896\,\text{s}$ that $Y = 0.22$. This is in approximate agreement with the more detailed computer codes.

In the remainder of this chapter we are going to be concerned with the sensitivity of this result to small variations of the parameters on which it depends. To this end we shall need to know how $X(T_F)$ varies with N_ν. Referring to (9.55) we see that

$$\frac{\delta b}{b} = -\frac{1}{2}\frac{\delta N_{\text{DF}}}{N_{\text{DF}}}. \tag{9.75}$$

Such a variation alters the scale of $K(y)$ and this produces a change in $X(T_F)$ given by

$$\frac{\delta X(T_F)}{X(T_F)} = \tfrac{1}{2}C\delta N_{\text{DF}}/N_{\text{DF}}, \tag{9.76}$$

where

$$C = \frac{1}{X(T_F)}\int_0^\infty dy' e^{y'} X_{\text{eq}}(y')^2 K(y') \exp[-K(y')]. \tag{9.77}$$

It is straightforward to evaluate this integral numerically. Doing so we find

$$C = 0.52.$$

Since

$$N_{\text{DF}} = 10.75$$

and since

$$\delta N_{\text{DF}} = \tfrac{7}{4}\delta N_\nu \tag{9.78}$$

we have that each new flavor produces a change in $X(T_F)$ given by

$$\delta X(T_F)/X(T_F) = 0.042. \tag{9.79}$$

We shall make use of this result shortly.

Given our assumptions, the cosmological helium production depends on four parameters. They are: N, the number of neutrino flavors; η, the baryon to photon ratio; τ_n, the neutron's mean life; and μ, a possible chemical potential for the electron neutrino. A chemical potential for the other neutrinos would be relevant only if it were large since then it would affect the energy density ρ. We will limit our discussion to small chemical potentials. The electron's chemical potential must be sensibly zero to preserve the charge neutrality of the universe. We want to study the variation in Y, δY, as a function of small variations in these parameters, variations away from their central values. We shall first give our final result for δY and then discuss its origin in some detail. As we shall see, the great advantage of having a semianalytic model for helium production is that these variations can be explicitly traced. Thus

$$\delta Y = 0.014\delta N_\nu + 0.18\frac{\delta t_n}{t_n} + 0.0044\frac{\delta\eta}{\eta} - 0.21\delta\mu, \tag{9.80}$$

where

$$\delta N_\nu = N - 3,\ \delta t_n = t_n - 896,\ \delta\eta = \eta - 5 \times 10^{-10}$$

and

$$\delta\mu = \mu.$$

We begin with a discussion of $\delta Y/\delta N_\nu$.

The variation of Y with respect to N_ν has two sources. The first we have already discussed. We showed that

$$\frac{\delta X(T_F)}{\delta N_\nu} = 0.042\, X(T_F). \tag{9.81}$$

The second source of the variation of Y with respect to N_ν comes about because the critical time t_c depends on N_ν since N_{eff}, which enters into the determination of t_c, depends on N_ν. Thus we can write

$$\begin{aligned}\frac{\delta Y}{\delta N_\nu} &= Y_0\left(0.042 - \frac{t_c}{t_n}\frac{\delta t_c}{\delta N_\nu}\right)\\ &= Y_0\left(0.042 - \frac{t_c}{t_n}\frac{\delta t_c}{\delta N_{eff}}\frac{\delta N_{eff}}{\delta N_\nu}\right)\\ &= Y_0\left(0.042 + \frac{t_c}{2t_n}\frac{1}{N^0_{eff}}\frac{\delta N_{eff}}{\delta N_\nu}\right).\end{aligned} \tag{9.82}$$

As we have seen, in this regime, to a very good approximation,

$$N_{eff} = 2 + (\tfrac{4}{11})^{\frac{4}{3}}\tfrac{7}{4}N_\nu. \tag{9.83}$$

Using $N^0_{eff} = 3.36$ and $t_c = 300$ s and $Y_0 = 0.22$ we find

$$\frac{\delta Y}{\delta N_\nu} = 0.014.$$

Finding $\delta Y/\delta t_n$ is entirely straightforward; i.e.,

$$\frac{\delta Y}{\delta t_n} = Y_0 \frac{t_c}{t_n^{\ 2}} = 0.18 \frac{\delta t_n}{t_n}. \tag{9.84}$$

The dependence of Y on η comes about because of the dependence of t_c on η. This, in turn, comes about via the Saha equation, (9.68), and the connection between time and temperature. Thus

$$\frac{\delta Y}{\delta \eta} = -Y_0 \frac{t_c}{t_n} \frac{\delta t_c}{\delta \eta}. \tag{9.85}$$

Using $N^0_{\text{eff}} = 3.36$ we have

$$\frac{\delta t_c}{\delta \eta} = -\frac{2.27\sqrt{(t_c/\text{s})^{\frac{1}{2}}}}{\epsilon_D/\text{MeV}} \frac{1}{\eta}. \tag{9.86}$$

Thus we find

$$\frac{\delta Y}{\delta \eta} = \frac{0.0044}{\eta}.$$

If it were correct to use X_{eq} in the helium production regime, finding $\delta Y/\delta\mu$ would be trivial. One would simply replace $\exp(\Delta m/T)$ by $\exp(\Delta m/T + \mu)$ in X_{eq} and then expand keeping only terms of lowest order in μ. Thus one would find, dropping relatively small terms,

$$X \simeq X_{eq}(1 - \mu) \tag{9.87}$$

and

$$\frac{\delta Y}{\delta \mu} \simeq -Y_0 \simeq -0.21.$$

However, it is not correct to use X_{eq} in this regime. It is at this point that we can take advantage of our semianalytical expression for X. We can expand in powers of μ within the integrals that lead to $X(T_F)$. Keeping the leading order in μ we find after expansion, using the previous notation that;

$$\begin{aligned} X(T_F) = X^0(T_F) + \mu\{X^0(T_F) - 2\int_0^\infty & e^{2y'}/(1+e^{y'})^3 \\ \times e^{-K(y')}\,dy' - \tfrac{1}{2}\int_0^\infty & K(y')e^{y'}1/(1+e^{y'})^2 e^{-K(y')}\,dy' \\ + b\int_0^\infty & (4/y'^3 + 1/y'^2) \times 1/(1+e^{y'})^2 e^{-K(y')}\,dy'\}. \end{aligned} \tag{9.88}$$

These one-dimensional integrals are readily performed. The second and third integrals are very small compared to the first, while the first integral is numerically about the same as the one occurring in the evaluation of $X(T_F)$ as the integrands are sensibly the same over most of the range of integration. Thus we find

$$X(T_F) \simeq X^0(T_F) - 0.149\mu \tag{9.89}$$

and thus

$$\frac{\delta Y}{\delta\mu} \simeq -0.21,$$

so that the numerical answer is about the same as the one obtained by using X_{eq} naively. Thus an additional flavor can be compensated by a μ of order 0.1.

Formulae for δY have appeared in the literature. They differ from ours in some respects. In the first place they seem to have been derived by fitting the numerically generated curves for Y, rather than from first principles. The coefficients for δN_ν and δt_n are, numerically, essentially the same as ours, but the coefficient for $\delta\eta$ one usually finds is some 2.5 times larger than ours. This can be traced to our use of (9.69) as the criterion for determining the onset of helium production. This criterion is qualitatively, but not quantatively, correct. The correct criterion can only be found by studying the rate equations for the capture and fusion reactions. (See Bernstein, Brown, and Feinberg, 1988, for a discussion in the spirit of this chapter.) When this criterion, which involves the capture and fusion rates, is used, the coefficient of the $\delta\eta/\eta$ term is increased to agree with the ones found in the literature, and the critical time is changed from 300 s to about 230 s. This has the effect of increasing Y by a few percent bringing our answer in close agreement with the computer code answers. To our knowledge no explicit analytic formula for $\delta Y/\delta\mu$ that goes beyond the one obtained by using X_{eq} has appeared in the literature.

Appendix A

In this appendix we wish to amplify the discussion following (3.66).[†] Let us use (3.16), (3.17), and (3.18) to write the Liouville operator in the form

$$L(f) = \left(p^\mu \frac{\partial}{\partial x^\mu} - \Gamma^\mu_{\alpha\beta} p^\alpha p^\beta \frac{\partial}{\partial p^\mu}\right) f. \tag{A1}$$

If we take for f the equilibrium function

$$f_{eq} = \exp[\alpha(t) + \beta^\mu(t) p_\mu], \tag{A2}$$

and demand that L annihilates this function then we find that

$$p^\mu \alpha_{,\mu} - \beta_{\alpha,\mu} p^\alpha p^\mu - \beta_\alpha \Gamma^\alpha_{\lambda\sigma} p^\lambda p^\sigma = 0. \tag{A3}$$

We may use this equation to deduce that, apart from special cases to be discussed,

$$(a)\ \alpha_{,\mu} = 0 \tag{A4}$$

and

$$(b)\ (\beta_{\lambda,\sigma} - \beta_\alpha \Gamma^\alpha_{\lambda\sigma}) p^\lambda p^\sigma = 0. \tag{A5}$$

We may now use the isotropy of space to reduce (A4) and (A5) to the case at hand. Thus, except in special cases,

$$(a)\ \dot{\alpha} = 0 \tag{A4$'$}$$

and

$$(b)\ \dot{\beta} p^0 p^0 - \beta \Gamma^0_{ij} p^i p^j = 0 = \dot{\beta} p^0 p^0 - \beta(\dot{R}/R) g_{ij} p^i p^j, \tag{A5$'$}$$

where, in the last step, we have used (1.13). For the $m = 0$ case

$$p^0 p^0 = g_{ij} p^i p^j, \tag{A6}$$

[†] I am grateful to J. L. Anderson and E. Weinberg for discussions concerning the subject matter of this appendix.

and (A5)′ implies that

$$\dot{\beta}/\beta = \dot{R}/R, \tag{A7}$$

and we recover

$$\beta = \text{constant} \times R. \tag{A8}$$

In general (A3) reduces to

$$\dot{\alpha} - \dot{\beta}p^0 + \beta \frac{\dot{R}}{R} \frac{p^i p^j}{p^0} g_{ij} = 0, \tag{A9}$$

and we recover (3.69). Equation (A5)′ is the Killing condition for a spatially constant β^μ, for $m \neq 0$. Hence the failure to have an equilibrium solution to the Boltzmann equation is related to the nonexistence of such a vector in general.

Appendix B

In this appendix[†] we would like to make explicit the connection between the conventional thermodynamics and the kinetic theory of the text. We begin with the Gibb's function G

$$G = U - TS + PV, \tag{B1}$$

where U and S are the total energy and entropy, P the pressure, and V the volume. G is to be considered a function of T, P, and N, the total number – conserved – of particles. As a function of N, G is assumed to have the property

$$G(T, P, N) = \frac{1}{\lambda} G(T, P, \lambda N), \tag{B2}$$

where λ is a numerical parameter. We can differentiate this expression with respect to λ and set $\lambda = 1$. Thus

$$0 = -G + \left.\frac{\partial G}{\partial N}\right|_{T,P} N, \tag{B3}$$

or

$$G = \left.\frac{\partial G}{\partial N}\right|_{T,P} N. \tag{B4}$$

On the other hand, from (B1)

$$\mathrm{d}G = \mathrm{d}U - T\mathrm{d}S - S\mathrm{d}T + P\mathrm{d}V + V\mathrm{d}P, \tag{B5}$$

and, from general thermodynamics,

$$\mathrm{d}V = T\mathrm{d}S - P\mathrm{d}V + \mu\mathrm{d}N, \tag{B6}$$

where μ is the chemical potential. Putting (B5) and (B6) together we have

$$\mathrm{d}G = -P\mathrm{d}V - S\mathrm{d}T + \mu\mathrm{d}N. \tag{B7}$$

† In some of the matters considered here I have had the pleasure of discussions with J. Peebles.

Thus

$$\left.\frac{\partial G}{\partial N}\right|_{T,P} = \mu. \tag{B8}$$

Therefore, from (B4), we have

$$G = \mu N = U - TS + PV. \tag{B9}$$

Thus we have derived the relation

$$S = (U + PV - \mu N)/T. \tag{B10}$$

From this derivation it is not clear how close to equilibrium we need to be for (B10) to remain valid. On the other hand, if we go to the statistical mechanics, i.e., Eq. (4.17), we see that in the language of (4.2) it is valid to $O(\phi^2)$. It is interesting to turn the argument around. Let us begin with the distribution

$$f_{\text{eq}} = e^{\alpha - \beta E}(1 + \phi), \tag{B11}$$

from which, using (4.14), we derive

$$S = V(\rho/T) + V[n(1 - \alpha)] + O(\phi^2). \tag{B12}$$

It is important to understand – see (4.29) – that α is, in general, a function of T. Now, keeping only terms of order ϕ,

$$\begin{aligned} dS &= \left.\frac{\partial S}{\partial V}\right|_T dV + \left.\frac{\partial S}{\partial T}\right|_V dT \\ &= \left[\frac{\rho}{T} + n(1 - \alpha)\right] dV + V\frac{\partial}{\partial T}\left[\frac{\rho}{T} + n(1 - \alpha)\right] dT \\ &= \frac{1}{T} d(\rho V) + \frac{P_0\, dV}{T} - \alpha n\, dV \\ &\quad + V dT \left\{\frac{\partial}{\partial T}[n(1 - \alpha)] - \frac{\rho}{T^2}\right\}. \end{aligned} \tag{B13}$$

In this equation we have used the approximately valid equation of state

$$P_0 = nT. \tag{B14}$$

If we evaluate the derivative in (B13) we arrive, to order ϕ^2, at the expression

$$dS = \frac{1}{T} d(\rho V) + P_0 \frac{dV}{T} - \alpha\, d(nV). \tag{B15}$$

We may now construct the conditions, for the expanding gas, under which dS/dt vanishes. Thus we demand that

$$\frac{dS}{dt} = \frac{1}{T}\frac{d}{dt}(\rho R^3) + n\frac{dR^3}{dt} - \alpha\frac{d(nR^3)}{dt} = 0. \tag{B16}$$

The quantity

$$\frac{d(nR^3)}{dt} = 0, \tag{B17}$$

by the conservation of particle number. The condition

$$\frac{1}{T}\frac{d}{dt}(\rho R^3) = -n\frac{dR^3}{dt} \tag{B18}$$

is just (4.32) in a slightly different guise; so we have succeeded in joining the statistical mechanics and the thermodynamics.

Appendix C

In this appendix we wish to discuss the question of the statistical mechanics of "light" neutrinos.† By "light neutrinos" we mean the neutrinos, and antineutrinos, that are actually observed in terrestrial laboratories, as opposed to hypothetical heavy neutral leptons which may, or may not, exist. As far as we know, there are three flavors of light neutrinos, ν_e, ν_μ, and ν_τ. There may be more flavors of which we do not, as yet, have direct evidence. As of this writing, no experiment forces us to assign a nonzero rest mass to any of these neutrinos. But experiment is consistent with the following upper limits

$$m_{\nu_\tau} < 70 \text{ MeV}, m_{\nu_\mu} < 250 \text{ keV}, m_{\nu_e} < 10 \text{ eV}.$$

The latter result has been the subject of the most diverse experimental study, ranging from the electron spectrum in tritium β-decay to an analysis of the arrival times of neutrinos from the supernova SN1987A. The 10 eV given above might be 20 eV and it might be zero. The question to be raised here is how many of these neutrinos are around now and what limits are thereby set on their putative masses?

To determine the number of light neutrinos presently around we must reconstruct their thermal history. Prior to their decoupling (see the discussion following (6.160)) the neutrinos will, apart from small corrections arising from their possible mass, be in an equilibrium distribution which, setting the chemical potentials equal to zero, we can write as

$$f_0 = \frac{1}{\exp[(m_\nu^2 + p^2)^{\frac{1}{2}}\beta_\nu] + 1}, \tag{C1}$$

where m_ν is the mass of any of the neutrinos and β_ν – which in this regime is essentially equal to β, the photon temperature – is the neutrino temperature. To simplify, we assume there is a $(\beta_\nu)_d$ at which the neutrino in

† There is extensive literature on this subject. See, for example, Szalay and Marx (1976), Cowsik and McClelland (1972), Bernstein and Feinberg (1981). Our treatment follows that of the last reference.

question decouples abruptly from the other leptons. For the purposes of the argument we are about to give, the details of the decoupling are unimportant. After the decoupling, the neutrino satisfies the Boltzmann equation with no collision term. We know from the discussion of Chapter 3 that any f of the form

$$f = f(Rp) \tag{C2}$$

will be a solution to the collisionless Boltzmann equation. Continuity suggests that the neutrino distribution, after decoupling, takes the form

$$f = \frac{1}{\exp[(m_\nu^2 + p^2R^2/R_d^2)^{\frac{1}{2}}(\beta_\nu)_d] + 1}. \tag{C3}$$

Strictly speaking, there is no temperature associated with this distribution, reflecting the fact that it is not an equilibrium distribution. Nonetheless, it contains, as we shall see, valuable scaling properties. We define the neutrino density, n_ν, as

$$\begin{aligned} n_\nu &= \int \frac{d^3p}{(2\pi)^3} \frac{1}{\exp[(m_\nu^2 + p^2R^2/R_d^2)^{\frac{1}{2}}(\beta_\nu)_d] + 1} \\ &= \frac{1}{R^3/R_d^3} \int \frac{d^3p'}{(2\pi)^3} \frac{1}{\exp[(m_\nu^2 + p'^2)^{\frac{1}{2}}(\beta_\nu)_d] + 1} \\ &= \frac{1}{R^3} R_d^3 (n_\nu)_d, \end{aligned} \tag{C4}$$

or

$$R^3 n_\nu = R_d^3 (n_\nu)_d, \tag{C5}$$

where the neutrino density at decoupling $(n_\nu)_d$ is given by

$$(n_\nu)_d = \int \frac{d^3p'}{(2\pi)^3} \frac{1}{\exp[(m_\nu^2 + p'^2)^{\frac{1}{2}}(\beta_\nu)_d] + 1}. \tag{C6}$$

We know from the considerations of Chapter 6, equation (6.173), that the photon temperature obeys the relation

$$\frac{T_<}{T_>} = \left(\frac{11}{4}\right)^{\frac{1}{3}} \frac{R_>}{R_<}, \tag{C7}$$

where "$<$" and "$>$" refer to "after" and "before" electron–positron annihiliation. The decoupling occurs before positron annihilation so, with the understanding that T is a temperature after annihilation, we can write

(C5) as

$$n_\nu = \frac{4}{11}\left(\frac{T}{T_d}\right)^3 (n_\nu)_d.$$

Unlike the zero-mass case, there are no simple scaling laws here for the average energy and momentum. We define the average energy, per neutrino, for example, as

$$\langle E\rangle_\nu = \frac{\displaystyle\int \frac{d^3p}{(2\pi)^3} \frac{(p^2+m_\nu{}^2)^{\frac{1}{2}}}{\exp[(m_\nu^2 + p^2R^2/R_d^2)^{\frac{1}{2}}(\beta_\nu)_d]+1}}{\displaystyle\int \frac{d^3p}{(2\pi)^3} \frac{1}{\exp[(m_\nu^2 + p^2R^2/R_d{}^2)^{\frac{1}{2}}(\beta_\nu)_d]+1}}$$

$$= \frac{\displaystyle\int \frac{d^3p'}{(2\pi)^3}(m_\nu{}^2 + p'^2R_d^2/R^2)^{\frac{1}{2}} \frac{1}{\exp[(m_\nu^2 + p'^2)^{\frac{1}{2}}(\beta_\nu)_d]+1}}{(n_\nu)_d}. \qquad \text{(C8)}$$

We shall be working in a regime in which

$$R_d/R \simeq 10^{-10} \ll 1,$$

and in which

$$m_\nu(\beta_\nu)_d \ll 1.$$

Hence, ignoring the relatively small effects of the quantum statistics, we can write

$$\langle E\rangle_\nu \simeq \frac{\displaystyle\int \frac{d^3p'}{(2\pi)^3}\left(m_\nu + \frac{R_d^2}{R^2}\frac{p'^2}{2m_\nu}\right)e^{-(\beta_\nu)_d p'}}{\displaystyle\int \frac{d^3p'}{(2\pi)^3} e^{-(\beta_\nu)_d p'}}$$

$$= m_\nu + 6 \times \left(\frac{4}{11}\right)^{\frac{2}{3}} \frac{T^2}{m_\nu} \equiv m_\nu + \langle \text{K.E.}\rangle. \qquad \text{(C9)}$$

We might define an "effective" neutrino temperature by equating

$$\langle \text{K.E.}\rangle \equiv \tfrac{3}{2} T^\nu_{\text{eff}}, \qquad \text{(C10)}$$

or

$$T^\nu_{\text{eff}} = 4 \times \left(\frac{4}{11}\right)^{\frac{2}{3}}\left(\frac{T}{m_\nu}\right) T. \qquad \text{(C11)}$$

This is to be compared to the neutrino temperature for massless neutrinos found in (6.175); i.e.,

$$T^{\nu} = \left(\frac{4}{11}\right)^{\frac{1}{3}} T. \tag{C12}$$

As the present photon temperature is $T \simeq 2.7$ K, which is in energy units $T \simeq 2.3 \times 10^{-4}$ eV, we see that for neutrinos in the electron-volt mass range $T^{\nu}_{\text{eff}} \simeq 10^{-3} T$. This assumes that the neutrinos have been expanding freely since T^{ν}_{d}. This might not be the case, if they are massive, since they could then undergo gravitational interactions which could change their distribution. For neutrinos in this mass range, the kinetic energy makes a negligible contribution to $\langle E \rangle_{\nu}$. Thus the neutrino contribution to the present energy density, assuming they are massive, is given by

$$\rho_{\nu} = \sum_{i} (n^{i}_{\nu} + n^{i}_{\bar{\nu}}) \mathrm{m}^{i}_{\nu}, \tag{C13}$$

where the sum is over flavors and we have exhibited both the ν and $\bar{\nu}$ contributions explicitly so as to be clear about the counting. From (C8)

$$n^{i}_{\nu} = \frac{4}{11}\left(\frac{T}{T_{\text{d}}}\right)^{3} (n^{i}_{\nu})_{\text{d}}. \tag{C14}$$

It is at this point that the significance of these neutrinos being "light" enters. We assume that they decouple when they are *relativistic*. The opposite limit is just the Lee–Weinberg problem treated in the text. We must then demonstrate that whatever mass limits we put on these light neutrinos are consistent with this assumption. The assumption is used to evaluate $(n^{i}_{\nu})_{\text{d}}$, which in this case is then simply proportional to $(n_{\gamma})_{\text{d}}$, the photon distribution evaluated at the same temperature. Thus, taking into account the fact that there are two states of polarization for the γ we have,

$$(n^{i}_{\nu})_{\text{d}} = (n^{i}_{\bar{\nu}})_{\text{d}} = \tfrac{3}{4} \times \tfrac{1}{2}(n_{\gamma})_{\text{d}}. \tag{C15}$$

The decoupling temperature is, because of the effects of the neutral currents, somewhat different for the various flavors. But, as the next step of the argument makes clear, this is irrelevant. The photon distribution scales as T^3. Thus we have, for the present neutrino distributions,

$$n^{i}_{\nu} = n^{i}_{\bar{\nu}} = \tfrac{4}{11} \times \tfrac{3}{4} \times \tfrac{1}{2} n_{\gamma} = 0.136\, n_{\gamma}, \tag{C16}$$

with $T = 2.7$ K, $n_{\gamma} \simeq 398/\text{cm}^3$, and thus†

$$n^{i}_{\nu} = n^{i}_{\bar{\nu}} = 54/\text{cm}^3, \tag{C17}$$

† This number density applies whether or not the neutrinos are Dirac or Majorana provided that the weak couplings are purely left-handed. At decoupling, according to hypothesis, the neutrinos are relativistic and the two theories with left-handed couplings are indistinguishable.

an interesting result in its own right. The dependence on flavor has canceled out. This result can be put directly into (C13) to find an expression for the neutrino contribution to the energy density. Thus

$$\rho_\nu = \sum_{\text{flavors}} m_i \times 108/\text{cm}^3. \tag{C18}$$

If we insist that this be less than ρ_c, as given by (6.183), i.e.,

$$\rho_c = \left(\frac{H}{75\ \text{km/s M}_{\text{pc}}}\right) \times 5.6 \times 10^{-3}\ \text{MeV/cm}^3, \tag{C19}$$

and, to get a feeling for the numbers, take

$$H = 75\ \text{km/s M}_{\text{pc}},$$

we have

$$\sum_{\text{flavors}} m_i < 52\ \text{eV}. \tag{C20}$$

If we are less generous with our assumptions we obtain a less stringent bound. To see how this works we need to recapitulate the standard view of the gross thermal history of the universe. So long as $T > m$, where m is the mass of any particle in thermal equilibrium with γs, the universe is "radiation" dominated, i.e., the Einstein equation is given by

$$\left(\frac{\dot{R}}{R}\right)^2 = \left(\frac{\dot{T}}{T}\right)^2 = AT^4, \tag{C21}$$

where

$$A = \frac{8\pi}{3} \frac{1}{M_{\text{pl}}^2} \frac{\pi^2}{30} N_{\text{DF}}, \tag{C22}$$

and

$$N_{\text{DF}} = \sum N_i^{\text{BE}} + \tfrac{7}{8} \sum_i N_i^{\text{FD}}. \tag{C23}$$

This equation can be solved to find the time as a function of temperature, or vice versa. Putting in the numbers, one has

$$t \simeq \frac{2.3}{(N_{\text{DF}})^{\frac{1}{2}}} \left(\frac{\text{MeV}}{T}\right)^2 \text{s}, \tag{C24}$$

which gives the "thermal clock" for the early universe. However, as the universe cools, the matter component of ρ becomes more and more important and, eventually, dominates. To see when this happens we solve the equation – see (4.28) and (4.31) –

$$n_{\text{B}} m_{\text{B}} \simeq 3 n_\gamma T, \tag{C25}$$

where n_B and m_B are the baryon number density and the proton mass, respectively. Hence the transition temperature is given by

$$T \simeq \frac{1}{3}\frac{n_B}{n_\gamma} m_B \simeq 3 \times 10^{-6} \text{ MeV}. \tag{C26}$$

Putting this into (C24) we see that the crossover temperature occurs at something like thirty thousand years. Hence if we solve the Einstein equation to find the lifetime of the universe, even for $k \neq 0$, it should be a good approximation to use the matter-dominated ρ throughout. Let us call the present age of the universe t_p and the present radius R_p, then

$$t_p = \int_0^{t_p} dt = \int_0^{R_p} dR/\dot{R} = \int_0^{R_p} \frac{dR}{[(\rho/\rho_c H_0^2 R^2) - k]^{\frac{1}{2}}}. \tag{C27}$$

In the last step we have used (2.18) and the definition

$$\rho_c = (3/8\pi) M_{pl}^2 H_0^2. \tag{C28}$$

To proceed, we wish to eliminate k in favor of more physically accessible observables. To this end, we may employ (2.8), i.e.,

$$\frac{k}{R^2} + \left(\frac{\dot{R}}{R}\right)^2 + 2\frac{\ddot{R}}{R} = -8\pi G_N P, \tag{C29}$$

where we have dropped the cosmological constant term. We next make an approximation which should be accurate in the regime in which nonrelativistic matter dominates ρ. What this matter is, is open to active experimental investigation. If there are relic neutrinos, any of whose masses are the order of 10 eV, then these will dominate ρ_p, the present ρ; or there might be nonshining hadronic matter – "Jupiters" – which dominate ρ_p. In any case, crudely speaking,

$$\rho_p \simeq nm, \tag{C30}$$

while

$$P \simeq nT. \tag{C31}$$

Since $T_p \ll m$, we shall drop P in (C29) and thus obtain an equation for k; i.e.,

$$k = -H^2R^2 - 2R\ddot{R}. \tag{C32}$$

But, in the absence of pressure, we can use (2.12) and (2.13) – Newton's law – to eliminate $R\ddot{R}$; i.e.,

$$\ddot{R} = -\tfrac{4}{3}\pi G_N R^2 \rho. \tag{C33}$$

Thus[†]

$$\begin{aligned} k &= -H^2R^2 + (8\pi/3)G_N R^2\rho \\ &= -H^2R^2 + H_0^2R^2\rho/\rho_c \\ &= -H_0^2R_p^2(1 - \rho_p/\rho_c). \end{aligned} \tag{C34}$$

The last equation follows because k is a constant which can be evaluated at $t = t_p$. Since, during this regime,

$$\rho \simeq nm \sim m/R^3, \tag{C35}$$

we have

$$\rho/\rho_p = (R_p/R)^3. \tag{C36}$$

Thus calling

$$x = R/R_p, \tag{C37}$$

we derive the expression,

$$H_0 t_p = \int_0^1 [1 - (\rho_p/\rho_c) + (\rho_p/\rho_c x)]^{-\frac{1}{2}} \mathrm{d}x \leqslant 1. \tag{C38}$$

The limiting case is for the empty universe. For the currently fashionable $k = 0$ case, i.e., $\rho_p = \rho_c$,

$$H_0 t_p = \tfrac{2}{3}. \tag{C39}$$

In general, if we suppose that massive neutrinos dominate ρ_p we can use (C38) to set limits on the neutrino mass providing we use input data for H_0 and t_p. Bernstein and Feinberg (1981) summarize this in a graph which we reproduce below. If we knew the neutrino masses, and if they dominated ρ_p, we could determine t_p, given H_0, or vice versa. (See Figure C.1.) Of course, the neutrinos may be sensibly massless and contribute negligibly to ρ_p.

The detection of these relic neutrinos would certainly be at least as exciting as the detection of the 3 K radiation. The problem is, of course,

† The reader should recall that $k = 0, \pm 1$ only in specially scaled units. Note also that it would be incorrect to neglect the k/R^2 terms in the Einstein equations since $\rho \sim 1/R^3$. In fact using (C34) we easily show that

$$\left|\frac{k/R^2}{\frac{8}{3}\pi G_N \rho}\right| = \frac{[1 - (\rho_p/\rho_c)]}{(\rho_p/\rho_c)}.$$

This ratio is, for $R = R_p$, greater than one unless $\rho_p = \rho_c$.

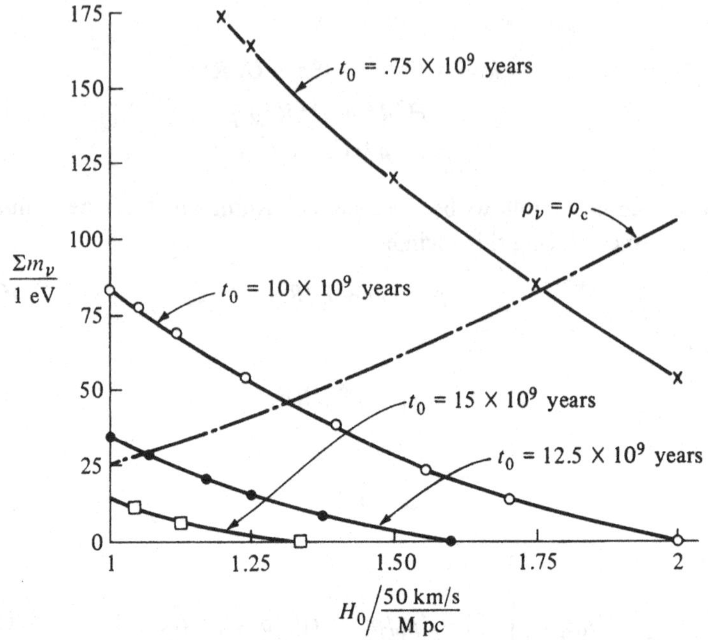

Figure C.1. The relation between the sum of neutrino masses, and Hubble's constant, for various present ages of the universe. The solid lines are the allowed values of $\sum m_\nu$ and H_0, for labeled values of t_0. The dot–dashed line indicates the values of $\sum m_\nu$, for a given H_0, at which $\rho_\nu = \rho_c$. From Bernstein and Feinberg (1981).

the very small weak interaction rate. However, in discussions of this rate it is sometimes overlooked that the neutrino mass can have a substantial effect. To see this, recall that Fermi's "Golden Rule" gives this rate as

$$w = 2\pi|T|^2\rho_f \simeq G_F^2 p_f^2 \frac{dp_f}{dE_f} dE_f \delta(E_f - E_i), \tag{C40}$$

where E_f and p_f are the final neutrino energies. If the neutrino is colliding from a nucleon of mass M then

$$E_f \simeq m_\nu + M + \frac{p_f^2}{2m_\nu} + \frac{p^2}{2M} \simeq M + \frac{p_f^2}{2m_\nu}. \tag{C41}$$

Hence

$$w \simeq G_F^2 m_\nu p_f \, dE_f \delta(E_f - E_i). \tag{C42}$$

If we use the approximate relation

$$p_f \simeq m_\nu v_i, \tag{C43}$$

where v_i is the incident neutrino speed, and note that, for massless neutrinos, this rate is approximately

$$w_0 \simeq G_f^2 E_i^2 c, \tag{C44}$$

we have

$$\frac{w}{w_0} \simeq \frac{m_\nu{}^2}{E_i^2}\frac{v}{c}. \tag{C45}$$

We can determine E_i from (C12); i.e.,

$$E_i \simeq 10^{-4}\ \text{eV}. \tag{C46}$$

To determine $\langle v \rangle$ we may use the scaling formula

$$\langle v \rangle = \frac{\int d^3p[p/E(p)]\exp[-(m^2 + p^2R^2/R_d{}^2)^{\frac{1}{2}}/T_d]}{\int d^3p \exp[-(m^2 + p^2R^2/R_d{}^2)^{\frac{1}{2}}/T_d]}$$

$$= \frac{R_d}{R}\frac{\int \frac{d^3pp}{[(R_d{}^2/R^2)p^2 + m^2]^{\frac{1}{2}}}\exp[-(p^2 + m^2)^{\frac{1}{2}}/T_d]}{\int d^3p \exp[-(p^2 + m^2)^{\frac{1}{2}}/T_d]}. \tag{C47}$$

Clearly as $m \to 0$, $\langle v \rangle \to 1$. For $R_d/R \ll 1$

$$\langle v \rangle \simeq \frac{R_d}{R}\frac{\langle p \rangle}{m} \simeq 3\frac{T_d}{m}\frac{R_d}{R}, \tag{C48}$$

where, to do the last step, we have evaluated (C47) in the $m \ll 1$ limit. Using (C7) we have, finally,

$$\langle v \rangle \simeq 3 \times \left(\frac{4}{11}\right)^{\frac{1}{3}} T/m. \tag{C49}$$

If we use, as an example, $m = 10$ eV, $T = 2.3 \times 10^{-4}$ eV, corresponding to $T = 2.7$ K, we find

$$\langle v \rangle / c \simeq 5 \times 10^{-5}. \tag{C50}$$

Hence, for this example,

$$w/w_0 \simeq 10^6 \tag{C51}$$

or

$$w \simeq 10^{-24}/\text{s}. \tag{C52}$$

It remains to be seen if an ingenious experimenter can make use of this tiny rate.

Appendix D

In this appendix we would like to expand on the analysis made of photon heating in the e^+–e^- regime. The principle tool of this extended analysis will be the relation

$$\frac{\partial U}{\partial t} = -3R^3P\frac{\dot{R}}{R}, \tag{D1}$$

where U and P are the *total* energies and pressures. The derivation of this relation, which follows from the Robertson–Walker equations, was given in Chapter 2. In the regime in question U consists of essentially three parts: U_ν, the neutrino part; U_γ, the photon part; and U_e, the e^+–e^- part. The first observation we wish to make is that during this regime, in which the neutrinos have decoupled, they obey (D1) *by themselves*. Since, during this regime, with T_ν the neutrino temperature,

$$RT_\nu = \text{constant}, \tag{D2}$$

we have

$$\frac{\dot{R}}{R} = -\frac{\dot{T}_\nu}{T_\nu}. \tag{D3}$$

Each neutrino degree of freedom in equilibrium, i, contributes to U_ν an amount

$$U_\nu^i = \frac{7\pi^2}{240}(RT_\nu)^3T_\nu. \tag{D4}$$

Thus, using (D3) and (D4) we have

$$\dot{U}_\nu^i = \frac{7\pi^2}{240}(RT_\nu)^3T_\nu\frac{\dot{T}_\nu}{T_\nu} = -U_\nu^i\frac{\dot{R}}{R}, \tag{D5}$$

and since, for a massless particle,

$$\tfrac{1}{3}R^3P = U, \tag{D6}$$

the result follows. Hence, as one might have expected, in this regime the relationship given by (D1) can be applied separately to the ν and the γ, e^+–e^- sectors. To use this relationship we must, of course, know U_ν and U_e which, in principle, means solving the system of coupled Boltzmann equations. The basic assumption we shall make is that the γs retain their equilibrium distribution throughout this period. Strictly speaking this is not true, but it should be a good approximation since the γs scatter strongly off the e^+–e^-. Hence we shall take, during this regime, with T the common e^+–e^-, γ temperature,

$$U_\gamma = \tfrac{1}{15}\pi^2 (RT)^3 T = \tfrac{1}{3}R^3 P_\gamma. \tag{D7}$$

There is no reason to take RT = constant during this regime. Hence, from (D7) we have

$$\dot{U}_\gamma = U_\gamma\left[3\frac{\dot{R}}{R} + 4\frac{\dot{T}}{T}\right]. \tag{D8}$$

Ignoring the small chemical potential, μ and introducing the pseudochemical potential α, we have, during this regime, for the e^+–e^- distributions $f_\pm$,

$$f_\pm \simeq \exp[-\alpha - \beta E]. \tag{D9}$$

Thus

$$n_\pm = 2e^{-\alpha(t)}\int e^{-\beta E}\frac{d^3p}{(2\pi)^3}, \tag{D10}$$

and for the e^+–e^- pressures, P_+, P_-,

$$P_\pm = Tn_\pm. \tag{D11}$$

Furthermore, calling N_1 the *total* e^+–e^- number we have in the $T < m$ regime,

$$U_1 = mN_1 + \tfrac{3}{2}TN_1. \tag{D12}$$

Hence,

$$\dot{U}_1 = \dot{N}_1(m + \tfrac{3}{2}T) + \tfrac{3}{2}\dot{T}N_1. \tag{D13}$$

Putting all of this together we find that (D1) implies that

$$4U_\gamma\left(\frac{\dot{R}}{R} + \frac{\dot{T}}{T}\right) = -\dot{N}_1(m + \tfrac{3}{2}T) - 3N_1T\left[\frac{\dot{R}}{R} + \tfrac{1}{2}\frac{\dot{T}}{T}\right]. \tag{D14}$$

This curious-looking expression has the correct limiting behavior. As N_1 and $\dot{N}_1$ tend to zero with the annihilation, the right-hand side vanishes and

$$TR = \text{constant}. \tag{D15}$$

On the other hand in the artificial situation with U_γ and $\dot{N}_l$ vanishing we have

$$TR^2 = \text{constant}, \tag{D16}$$

the equilibrium temperature distribution for the massive l particles. For our purposes we want to collect terms in $\dot{T}/T$. Thus

$$\frac{\dot{T}}{T} = \frac{-\dot{N}_l(m + \frac{3}{2}T) - \dfrac{\dot{R}}{R}(4U_\gamma + 3N_l T)}{4U_\gamma + \frac{3}{2}N_l T}. \tag{D17}$$

Since $\dot{N}_l < 0$ and $\dot{R}/R > 0$, $\dot{T}/T$ could, in principle, have either sign. However, the physics of the situation would suggest the following: Presumably

$$|\dot{N}_l| \sim N_l \langle \sigma v \rangle N_l.$$

Hence, so long as the annihilation rate dominates $\dot{R}/R$, we expect the temperature to *increase*. But, as N_l is a rapidly decreasing function of T the e^+–e^- rapidly disappear and $\dot{T}/T$ then becomes negative. During the initial epoch in which e^+–e^- annihilation and recombination are comparable, one expects to have $\dot{N}_l < \dot{R}/R$. Hence there will be, in general, a transient phase during e^+–e^- annihilation in which the γs will cool before the heating begins.

Appendix E

In Chapter 6 we presented a very accurate analytic solution to the rate equation – (6.104). Implicit in this solution was the assumption that the parameter $\hat{\lambda}$, which enters the equation, is independent of temperature. This is true in the examples we selected, but there are examples, which have been considered in the literature, where this is not true. We will therefore drop this condition.[†] The equation to be solved is now,

$$\frac{\mathrm{d}\hat{G}(y)}{\mathrm{d}y} = \hat{\lambda}(y)[\hat{G}(y)^2 - \hat{G}_0(y)^2]. \tag{E1}$$

We introduce the variable

$$\phi = \phi(y) = \frac{1}{\hat{\lambda}_0}\int_{y_0}^{y} \mathrm{d}y'\hat{\lambda}(y'), \tag{E2}$$

where we will specify $\hat{\lambda}_0$ and y_0 shortly. Thus (E2) becomes, using the transformation,

$$\hat{G}(y) = \frac{1}{\hat{\lambda}_0 f(\phi)}\frac{\mathrm{d}f(\phi)}{\mathrm{d}\phi}, \tag{E3}$$

$$\left(\frac{\mathrm{d}^2}{\mathrm{d}\phi^2} - \hat{\lambda}^2\hat{G}_0[y(\phi)]^2\right)f(y) = 0, \tag{E4}$$

which is the present equivalent of (6.107). Hence, as in Chapter 6, we can use the WKB approximation above the transition temperature. Thus, in this region,

$$f(\phi(y)) \simeq \hat{G}_0(y)^{-\frac{1}{2}}\exp\left[-\int^{y}\mathrm{d}y'\hat{\lambda}(y')\hat{G}_0(y')\right], \tag{E5}$$

and

$$\hat{G}(y) \simeq \hat{G}_0(y) + \frac{\hat{G}_0(y)}{2\hat{\lambda}(y)\hat{G}_0(y)}. \tag{E6}$$

[†] This treatment was done in collaboration with L. Brown and G. Feinberg.

The transition point y_0 is given by

$$\frac{\pi\hat{\lambda}(y_0)^2}{8\xi(3)^2}\, y_0 e^{-2/y_0} = 1. \tag{E7}$$

We take y_0 to be the lower limit of the integral in (E2). We choose

$$\hat{\lambda}_0 = \hat{\lambda}(y_0), \tag{E8}$$

so that

$$\phi \simeq y - y_0, \tag{E9}$$

near y_0. Since $\hat{\lambda}(y_0)$ is, in general, large

$$\ln \hat{\lambda}(y_0) \simeq 1/y_0 \gg 1. \tag{E10}$$

The extrapolation across the $y \simeq y_0$ region is determined by the equation

$$\left[\frac{d^2}{d\phi^2} - \frac{1}{y_0^4}\exp\left(\frac{2\phi}{y_0^2}\right)\right] f(\phi) = 0, \tag{E11}$$

where we have used (E6) and (E8). This equation has the solution

$$f(\phi) = K_0(e^{\phi/y_0^2}). \tag{E12}$$

As $1/y_0 \gg 1$ in these applications, we can use the small argument asymptotic form of the Hankel function. Thus

$$f(\phi) \simeq -(\phi/y_0^2) + \ln 2 - \gamma. \tag{E13}$$

Hence

$$\hat{G}(0)^{-1} \simeq \int_0^{y_0} dy' \hat{\lambda}(y') + \hat{\lambda}_0 y_0^2(\ln 2 - \gamma). \tag{E14}$$

Since $y_0 \ll 1$ we have approximately

$$\hat{G}(0)^{-1} \simeq \int_0^{y_0} dy' \hat{\lambda}(y'), \tag{E15}$$

where y_0 is determined by (E7). When $\hat{\lambda}$ is independent of y we reduce to the case considered in Chapter 6.

To get an idea of how large an effect this y dependence is, let us consider an example that has been discussed in the literature.† This example also involves massive L particles annihilating into l particles. However, here the

† See Krauss (1983). The treatment is not done correctly, since the relevant rate equation is not correctly solved.

L particles are taken to be Majorana. This means that $\mathrm{L} = \bar{\mathrm{L}}$. It turns out† that this implies that the annihilation is in *p*-waves. This implies, in turn, that the rate equation takes the form

$$\frac{\mathrm{d}\hat{G}(y)}{\mathrm{d}y} = \hat{\lambda} y[\hat{G}(y)^2 - \hat{G}_0(y)^2], \tag{E16}$$

where $\hat{\lambda}$ is essentially the same as the $\hat{\lambda}$ of (6.105). If we apply (E15) we see that in this case

$$\hat{G}(0) \simeq \frac{1}{\frac{1}{2}\hat{\lambda} y_0^2} \simeq 2 \ln \hat{\lambda} \frac{\ln \hat{\lambda}}{\hat{\lambda}}. \tag{E17}$$

We see that this differs from the result we would have obtained by taking $\hat{\lambda}$ independent of y by a factor of $2 \ln \hat{\lambda}$. Since, in the problem at hand, $\hat{\lambda} > 10$ this can be a very substantial effect. The reason that more L particles remain, asymptotically, in this case is that the annihilation is suppressed at low energies due to the *p*-wave character of the interaction.

† See Krauss (1983).

References

Anderson, J. L. (1970) in *Relativity*, Carmeli, M., Fickler, S. I., and Witten, L. (eds.), Plenum Press, p. 109.

Bergsma, F. *et al.* (1983) *Phys. Lett.*, **128B**, 361.

Bernstein, J., and Feinberg, G. (1981) *Phys. Lett.*, **101B**, 39.

Bernstein, J., Brown, L. S., and Feinberg, G. (1985) *Phys. Rev.*, **D32**, 3261.

Bernstein, J., Brown, L. S., and Feinberg, G. (1988) *Rev. Mod. Phys.* (in press).

Bethe, H. A., and Salpeter, E. E. (1957) *Quantum Mechanics of One- and Two-Electron Atoms*, Academic Press, New York.

Callan, C. G., Dicke, R. H., and Peebles, P. J. E. (1965) *Am. J. Phys.*, **33**, 105.

Cowsik, R., and McClelland, J. (1972) *Phys. Rev. Lett.*, **29**, 669.

Dicus, D., Kolb, E., and Teplitz, V. (1977) *Phys. Rev. Lett.*, **39**, 168.

Dicus, D., Kolb, E., and Teplitz, V. (1978) *A. P. J.*, 221.

Ehlers, J. (1971) in *Proceedings of the Varenna Summer School on Relativistic Astrophysics*, Academic Press, New York.

Einstein, A. (1917) *Sitzungsberichte der Preussichen Akad. d. Wiss.*, **142.**

Fackler, O., and Tran, Thanh Van (eds.) (1986) *'86 Massive Neutrinos*, Editions Frontieres.

Guth, A. H. (1981) *Phys. Rev.*, **D23**, 347.

Harari, H., and Nir, Y. (1987) *Nuc Phys.*, **B292**, 231.

Israel, W., and Vardalas, J. N. (1970) *Lett. al. Nuov. Cim.*, **4**, 887.

Kamponeets, A. S. (1957) *J. E. T. P.*, **4**, 730.

Kolb, E. W., and Olive, K. A. (1986) *Phys. Rev.*, **D33**, 1202.

Krauss, L. L. (1983) *Phys. Lett.*, **128B**, 37.

Lee, B. W., and Weinberg, S. (1977) *Phys. Rev. Lett.*, **39**, 169.

Lemaître, G. (1925) *Journal of Math. and Phys.*, **4**, 188.

Peebles, P. J. E. (1966) *A. P. J.*, **146**, 542.

Peebles, P. J. E. (1971) *Physical Cosmology*, Princeton University Press, Princeton.

Preskill, J. P. (1979) *Phys. Rev. Lett.*, **43**, 1365.

Robertson, H. P. (1929) *Proc. Nat. Acad. Sci.*, **15**, 822.

Sarkar, S. (1986) "Superstrings...," eds. G. Furlan *et al.*

Scherrer, R. J., and Turner, M. S. (1985) *Phys. Rev.*, **D31**, 681.

Scherrer, R. J., and Turner, M. S. (1986) *Phys. Rev.*, **D33**, 1585.

Schücking, E. L., and Spiegel, E. A. (1970) *Comments Ap. and Space Phys.*, **2.**

Sitter de, W. (1917) *Monthly Notices R. A. S.*, **78**, 3.

Stewart, J. M. (1971) *Non-equilibrium Relativistic Kinetic Theory*, Springer-Verlag, New York.

Szalay, A. S., and Marx, G. (1976) *Astron. Astrophys.*, **49**, 437.

Vysotskii, M. I., Dolgov, A. D., and Zel'dovich, Ya B. (1977) *JETP Lett.*, **26**, 188.

Weinberg, S. (1971) *A. P. J.*, **168**, 175.

Weinberg, S. (1972) *Gravitation and Cosmology*, John Wiley and Sons, New York.

Zel'dovich, Ya B. (1965) *Adv. Astron. Astrophys.*, **3**, 241.

Zel'dovich, Ya B., Kurt, V. G., and Sunyaev, R. A. (1969) *J. E. T. P.*, **28**, 146.

Zel'dovich, Ya B., and Sunyaev, R. A. (1969) *Astrophysics and Space Science*, **4**, 301.

Zel'dovich, Ya B., and Khlopov, M. Yu (1978) *Phys. Lett.*, **79B**, 239.

Index

Printed in the United States
By Bookmasters